T0292405

Teaching Practical Theatrical 3D Printing

Teaching Practical Theatrical 3D Printing: Creating Props for Production is a cohesive and practical guide for instructors teaching 3D printing techniques in stagecraft, costume and props courses.

Written for the instructor, this book uses non-technical language to explain 3D printers, their workflows and products. Coverage includes the ins and outs of multiple filaments, pros and cons of different types of printers, shop or laboratory setup and safety concerns. The book features lesson plans, rubrics and class-tested sample student projects from design to finished product that highlight learning objectives and methodologies, as well as software and hardware usage explanations and common problems that can occur within design and printing. Step-by-step instructions are included for many types of projects, including fake noses, candlestick phones, buttons, 3D scans, historical recreations and linear actuators. The book also contains examples of poor, average and excellent work with grading explanations and guidance on how to help the student move to the next level with their projects. Chapter objectives, chapter summaries, checklists and reflection points facilitate an instructor in gaining confidence with 3D printers and incorporating their use in the classroom.

Teaching Practical Theatrical 3D Printing is an excellent resource for instructors of Props and Costume Design and Construction courses that are interested in using state of the art tools and technology for theatre production.

Fully editable files for every object featured in the book are available at www.routledge.com/9781032453279, allowing readers to jump-start their projects and giving them the flexibility to change and redesign the items to best fit their needs.

Robert C. Berls is Professor of Drama at UNC Asheville. He is a professional scenic, lighting, properties, costume and sound designer with 12 years of experience in rapid prototyping and 3D printing. A member of USITT (United States Institute for Theatre Technology), he was the vice commissioner for Health and Safety for the Scene Design Commission and co-leader of the national Tech Olympics competition.

Teaching Practical Theatrical 3D Printing

Creating Props for Production

Robert C. Berls

Routledge
Taylor & Francis Group

NEW YORK AND LONDON

Designed cover image: Robert C. Berls

First published 2024
by Routledge
605 Third Avenue, New York, NY 10158

and by Routledge
4 Park Square, Milton Park, Abingdon, Oxon, OX14 4RN

Routledge is an imprint of the Taylor & Francis Group, an informa business

ISBN: 978-1-032-45331-6 (hbk)
ISBN: 978-1-032-45327-9 (pbk)
ISBN: 978-1-003-37648-4 (ebk)

DOI: 10.4324/9781003376484

Typeset in Stempel Garamond LT Pro
by Newgen Publishing UK

Access the Support Material: www.routledge.com/9781032453279

For Dorshorst, Nelson, and Davis.
To all those in my life and in the life of others who fostered
curiosity and embrace the asking of "why", thank you.

Contents

1. How to Fuse your Filament   1
    X Y Z and Me, or the Why and How of a
    3D printer   *1*
    Slicing and Build   *4*
    Brims, Rafts and Supports   *6*
    Is it all Plastic?   *8*
    Raw Materials   *12*
    Finishing   *15*
    Printer Selection   *16*
    Common Problems and Corrections   *19*

2. Teaching and Problem Solving   23
    The What, Where, When and Why of How   *25*
    Rigor Based on Attempt and Process Instead
    of Product   *26*
    Games and Competition   *27*
    Individual vs Group   *28*
    The Process of the Process Lab   *30*

3. Level 1: Lettered Card   33
    The Project   *38*

4. Level 2: Candlestick Phone: Rubric,
    Methods, Ways and Means   48
    Additional Rigor   *56*
    Challenge Areas in Brief   *59*

5. Level 3: Scanning a Real-World Item   60
   CPSG 3D Scanner   *64*
   Additional Rigor   *65*
   Challenge Areas in Brief   *68*

6. Level 4: Makeup Prosthesis: Rubric,
   Methods, Ways and Means   70
   Additional Rigor   *78*
   Challenge Areas in Brief   *80*

7. Moving Parts   81
   Problems Solving with Electrical Connections   *86*
   Additional Rigor   *87*
   Challenge Areas in Brief   *89*

8. Large Prints Using Small Printers   90

9. Additional Projects   102
   Science Fiction Communicator   *102*
   Museum Replica   *104*
   Jewelry   *107*
   Tools   *110*
   Clamps   *110*
   Measuring Aids   *111*
   Centering Jigs   *114*

10. Sum Up   117

11. Glossary of Terms   120

Index   *123*

1

HOW TO FUSE YOUR FILAMENT

X Y Z and Me, or the Why and How of a 3D printer

No matter how many times I teach a class on the construction and design of 3D printers this question, the how of 3D printing, is pervasive and needs answered in a basic way. These basics are the foundation for understanding the rest of the process. The answer is, according to most of my theatre students, *magic*. Just because you have selected the arts as your degree or passion does not diminish that fact that math is the core and language of most of the universe including you producing a 3D item that is fun to look at and play with.

Let us divide a space, in this instance, a cube that is roughly eight inches by eight inches by eight inches (Figure 1.1).

DOI: 10.4324/9781003376484-1

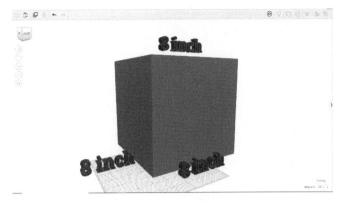

Figure 1.1 Cube that is 8 x 8 x 8 inches or 512 cubic inches

Five Hundred and twelve cubic inches that fills with plastic to make your day brighter. The x-axis moves left to right, the y-axis moves forward and backward, and the z-axis moves up and down. This is where stepper motors come in. These dual phase motors, having an A side and a B side to the wiring, have 200 steps or positive stops per full rotation of the shaft if they are 1.8-degree motors. Each step represents 1.8 degrees of rotation. If you take a non-energised stepper motor and turn it with your fingers, you will feel the stops as you turn. These turns controlled with a controller, in our case an Arduino, Raspberry Pi, or proprietary control card that has stepper motor drivers attached an example being an Arduino mega 256 with added 3D printer shield), turn the motor the 200 steps or with use of digital manipulation 3,200 steps per rotation. The card takes the item you put into your interface program and pulses the motor incrementally to gain resolution within your print.

The belts and threaded rods translate those pulses into movements in the x, y, and z-axis so that it fabricates a model. It all comes down to how many pulses equal a millimeter when motion through the belts and threaded rods transmits. It is fine-tuning of steps per millimeter that poses a challenge but with software conversion and simple algebra, it becomes possible. There are also many other sizes of stepper motors available including larger Nema 23 and 34 and smaller Nema 14 and 11. All of these steppers have different specifications and uses and may not work well for 3D printing (Figure 1.2).

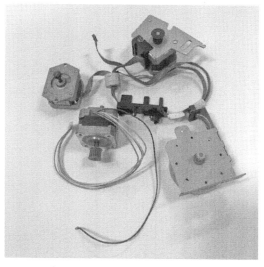

Figure 1.2 Used stepper motors

Most 2D printers today use a simple DC motor with an encoder. These are fine for moving a lightweight print head but lack the resolution and power to move 3D printing components simply and efficiently. I am sure that someone will cleverly use this type of motor to build a successful printer, but this level of experimentation surpasses this book.

To sum up, print resolution in part, comes from stepper motors that are baby stepped through hardware and software.

Slicing and Build

After creation of a 3D virtual item, it is time to turn it into reality through slicing. Imagine an orange. You want to recreate said citrus item in 3D reality. You have modeled the fruit in your favorite modeling program. Now it is time to build your genius. This is where a slicing program comes in. Just like the orange, you have your choice of flavors. Cura is a tart yet sweet mix of settings, slic3r is a formidable standard that most like but is no longer being supported. Each maker will prefer different formats and settings and think that their choice is the best choice. Ultimately, every program will break up your orange into 0.1 to 0.3mm layers that stack one on top of the other and use this to create your citrus perfection in plastic.

The first layer is the most critical. You need the proper amount of "squish" on the plate to gain adherence so that the next layer does not pull the fledgling model off the build surface. Temperature of the extruding filament and heated bed is essential to gain this hold. When in doubt use a glue stick on the build plate. I will sometimes have to adjust a printer manually when it is in its warmup after homing when using a Prusa clone printer. Accomplish this by rotating the z-axis clockwise or counterclockwise to get the desired height. I like to run my finger over the first few lines to make sure they have good adhesion. Visual inspection of the first layer can also show that "squish" has been achieved. For other printers, you will either have to level the print bed, adjust your z-axis measurements above the build plate, etc. Follow your manufacturers trouble shooting chart or process for adjusting the z-axis (Figure 1.3).

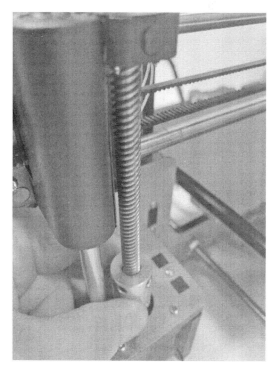

Figure 1.3 Hand z-axis adjustment

When this is accomplished, your print has a higher level of success. With good "squish," you can move onto the next level of your print.

Brims, Rafts and Supports

Brims and rafts are structures that help with the stability of the print on the plate and can provide better adhesion for the print. These settings are usually found in the slicing manger section of your software and can be toggled on or off before slicing. Consider using these support structure for detailed prints. Brims are an outline around the base of the print and happen before the first layer extrusion of the print. Consider a brim the first run to make sure your hot end is extruding and to work out any old filament that has taken up residence on the nozzle tip. Brims also help you to check your "squish" and hold down any surface you are printing on. In the old days of last week, makers would use painters' tape as a printing surface. The brim would make sure your tape was down in the model area.

Consider a raft a footer or foundation to your model. The printer will lay down a grid on the build plate before printing the first layer of the model. It will provide a steady base and good adhesion point for your model, especially if it has a narrow base that widens out as it prints. After the print is completed, remove the raft from the bottom (Figures 1.4a and 1.4b).

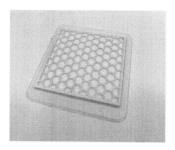

Figure 1.4a Raft

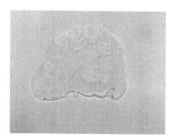

Figure 1.4b Brim

Supports are exactly what they sound like. Any over-hang generates supports, a removable structure. You can choose touching bed, overhang support or full support. These will increase print times but can increase the potential of a successful print. Not all structures will remove as easily as others, so be aware of what your preferences are and use them consistently (Figure 1.5).

Figure 1.5 Support structure

We will be focusing on fused filament fabrication due to equipment cost and raw materials. I have found resin prints reached a higher detail per print, but expense, curing issues and skin sensitivity to resins make it less useful for a beginning to intermediate level of prop 3D printing. PLA filament is easy to use, easy to glue together and manipulate after it is printed.

Is it all Plastic?

The short answer is no. The long answer is also no. There are so many variables to density and infill that, even when meaning to print all plastic, the item will have voids and gaps. That makes it not 100% plastic. Even at 100%, the complete integration of stranded material will not create the same product as injection molding will without post processing. Props makers through the years have concluded that ultimate strength to time of print achieves itself at 50% infill. As infill goes above 50%, the chance of warping and print failure on fused filament printer's increase as well as time for the print and material usage. Functional prints, prints that will support weight, stress, etc., can go above 50% but tests should be in your own experiments, 30% infill is typically the maximum I will use for hand props. This is just an average for the entire print. I will sometimes vary the infill density in different areas of the print. Maybe the first 20% of the print is 15% and then the rest is 45% due to being the handle of the item. The average of 30% for the entire print, a mix of speed of print and overall strength, make this the "sweet spot" for maximum effectiveness.

The shape of the infill is also critical for different applications. What infill will create the best result for your prop? Will it provide the correct support for handling and usage of the item? Let us break down the three infills I use on a regular basis as a starting point. The first is the grid shape of infill.

The grid is my go-to infill for the mix of speed and strength. It creates a series of interconnecting lines, creating it on the bias, drawn on the 45-degree angle to the main axis of the item. This interconnected grid provides a comfortable level of support and crush resistance for most items (Figure 1.6).

Figure 1.6 Grid infill

This pattern creates channels that run from top to bottom of a print. Insertion of dowels, resin or any type of reinforcement material takes place to add strength or weight to the print. This will require preplanning and a printing pause for insertion during the print. Make sure to have your reinforcement's material below the print level before resuming your print (Figure 1.7).

Figure 1.7 Grid infill with poured in resin

The second most used infill is hexagon, which creates a honeycomb pattern that also goes from top to bottom of the print. This infill pattern is not natively supported by some slicing software including Cura. The line shape change and increase in length of the line increase the amount of filament used and time that each layer takes to print. I like to use hexagon to give more sidewall strength to the item.

Will your prop be dropped in the course of using it? Does it need more damage resilience or be able to be stepped on, etc.? Hexagon infill might give you the best result (Figure 1.8).

Figure 1.8 Hexagon infill damage resistance

The third basic infill is a series of parallel lines called line. The line infill gives strength on two sides of the print and flexibility to the other two sides.

This is probably the fastest infill pattern but you will sacrifice ultimate strength for speed (Figure 1.9).

Figure 1.9 Line infill

Now we will discuss the raw materials of the printing process. This raw material comes in many forms: filament, liquid, paste, powder, etc.

Raw Materials

We will first start by examining filament. Filament, or a continuous strand of homogenous material, typically comes in two sizes depending on your printer: 1.75mm and 3.0mm. This filament pushed down and extrudes into a strand that is 0.3mm, 0.4mm, etc., from the hot end. The size differences are accomplished by changing the nozzle size instead of software settings. This strand will form your item. My go-to material for general prop construction is PLA or polylactic acid. PLA is great and easy to work with but is less durable and is suspectable to impact so using it for long runs of more than ten shows might not be recommended. There are also several other materials available with new mixes coming out every month. Fused filament fabrication is the focus of this discussion due to the cost-effective nature and easy to use methodology.

Material	Usage	Temperatures	Pros and cons	Safety
Polylactic acid (PLA)	General use material that is easy to print and lower cost than most filaments.	Hot end: 210°C Bed: 50°C	Easy to print but is susceptible to water absorption from atmosphere. Susceptible to warping when printing and is not very impact resistant.	Biodegradable. Typically, does not need specialised fume evac other than room ventilation.
Acrylonitrile butadiene styrene (ABS)	Impact and heat resistant material that is good for high use applications.	Hot end: 225–245°C Bed: 90–110°C	Heat and impact resistant plastic that is good for interior and exterior applications. Can warp dramatically due to lack of bed adhesion. Susceptible to water absorption from atmosphere.	Strong order when printing. Advisable to have fume evac within interior printing facilities. Check SDS of specific filament for safety recommendations.

(Continued)

Material	Usage	Temperatures	Pros and cons	Safety
Glycolised polyester (PETG)	Combines the easy printing of PLA with the strength of ABS.	Hot end: 220–255°C Bed: 60–80°C	Food compatible and chemical resistant plastic that is susceptible to water absorption from the atmosphere and extreme warping due to bed adhesion issues.	Like PLA, PETG typically does not need specialised fume evac but evaluate your specific space for needs.
Nylon (polyamide) (extrusion temperature cannot usually be achieved in entry level printers)	A strong impact resistant plastic that can be finicky.	Hot end: 220–250°C Bed: you will have to dial it in for best results.	Really strong but is susceptible to warpage.	This filament will require full venting of fumes. Can cause headache, burning eyes, etc., and potentially releases harmful VOCs.
Flexible (TPU) (soft PLA)	Flexible material that is rubber-like and quite durable.	Hot end: check manufacturers recommendations for temperatures.	Can make belting molds, etc.	Check the safety data sheets of different manufactures for safety data.

It is always important to check the manufacturer's safety data sheet (SDS) for complete handling and safety protocols for usage as not all PLA is completely PLA. The main items to check are product composition, if they produce ultra-fine particles, and when extruded do they produce volatile organic compounds (VOCs) that you can breathe in. Do you need HEPA filtration or carbon filters? Can this be accomplished through an add-on or do you need to have extraction to a hood setup? Usually the more exotic the material, the more safety protocols must be enacted. When in doubt, vent it out – safely.

Finishing

The last step before painting is finishing the print. This happens in several ways. First, is mechanical or sanding. Using a 220-grit sandpaper, a smooth uniform finish can be attained after careful sanding. Using a power sander is not recommended due to the buildup of heat. This can melt your print and mean starting all over again. Continue to up your grit of paper (possibly up 2,000) to attain a factory finish. This can be time consuming and create a lot of dust. Please use the appropriate PPE and consider the space where finishing will take place. Another mechanical method would be to use either walnut shell or corncob in a vibratory tumbler to finish the print. This is a "set-it-and-forget-it" method. Noise and space will be the only drawbacks to using a tumbler.

Second, will be chemical finishing. Depending on the filament, acetone, alcohol, etc., smooths the outer surface of the print. For ABS filament, vaporising acetone in a closed container with heat and placing the print inside is a preferred method for some makers. This can be dangerous and highly flammable if handled incorrectly. Please use appropriate caution and PPE if using chemical finishing.

The third method is coatings. Coatings, including paint, can finish the exterior of a print, but careful selection of materials is essential.

The fourth method is heat. Some makers place their prints in a toaster oven at 160–170°C in order fuse the external layers into a uniform plastic finish. Again, depending on the filament, this could release harmful fumes and could cause a fire within the oven cavity.

Printer Selection

This brings us to selecting the printer that will serve our needs completely, or at least well. The printers form into categories based on the budget safety time (BST) triangle (Figure 1.10).

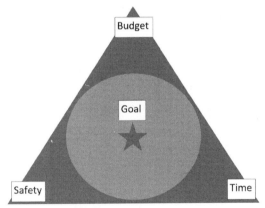

Figure 1.10 Budget, safety, time triangle

I hope that we can get a balanced printer that we can afford that will serve our making needs. I will be using common printer names to identify most of the printers you will see for purchase and explain benefits and drawbacks as well as balance them in the idea of the BST triangle.

One of the lowest cost fused filament fabrication (FFF) 3D printers is the gantry x-axis with the print head adjusting height on the z-axis. These printers go by multiple names including Mendel, Prusa or bed y-axis. As the last name implies, the print bed moves on the y-axis and the print head moves on the x-axis and raises or lowers the height in the z-axis. Commerical printers like Anycubic, Creality, Ultimaker and Makerbot offer many of the same features as well as additional functionality in a non-kit format. Using between four and five stepper motors, these printers have a lower mechanical part count as well as using lower cost 8mm or 5/16" steel rod for kinematics. Depending on the complexity of the device and size of the print area, these printers can retail from $100 to $1,000 for an assemble-your-own kit. I have found that this is a great starter design for most makers and will teach you the maintenance lessons that will serve most makers in the future. You will have to tighten and check before each major print or at least every 10–20 hours of printing time.

Another type of printer that upgrades print quality by reducing excess vibration is the core xy printer. This uses the same number of stepper motors but controls the z-axis by raising or lowering the bed, which does not move on any other axis. The print head moves in a structure that slides on the x-plane. This acts to minimise excess vibration by stabilising the heavier print head in a 2D-plane. Typically, these printers will cost from $300 and up and will usually have a closable casing to protect against air gusts and house the belting that transfers power from the motors to the physical structures.

Another printer that is rapidly scalable in the z-axis is a Kossel or Rostock printer. These printers use three stepper motors and towers to move an interconnected print head in three dimensions. A lower part count and a smaller needed support structure can make these printers affordable. They can be susceptible to vibration errors and slower print speeds. Highly scalable but can be prone to excess print failure and slow speed.

The last printer I will cover in the affordable category is resin cure printers. These printers do not use filament, but a photo curing liquid that prints upside down. These resin printers can produce amazing detail and structure quickly, but have a higher material cost due to the resin. They also produce a more fragile print that cannot be joined with heat or manipulated easily to change the design. The software that controls these printers is different than the slicing style software for FFF printers. You will not be able to create infill as you would in FFF printing. It will either be solid or have varying levels of hollowness. Remember to include a drain in the design so that the interior will drain the unused resin. The resin can sometimes cause skin sensitivity when handled in liquid form and will typically need a UV curing chamber for post processing.

The question then becomes how much you can afford, how fast does it need to produce and what safety protocols can or need to be implemented? My go-to printer is the gantry x printer. There is more maintenance that needs time to accomplish but cost to production ratio makes it a strong choice.

Common Problems and Corrections

Not all prints will succeed. There are any number of problems and solutions to make your print fail or succeed. Let us break these problems down into three categories: adhesion, extrusion and mechanical.

Adhesion means the filament sticking to the bed and to each layer. If the filament does not adequately stick to the bed, the corners of the item can pull up and create an uneven based or squeezed edge. This fixes itself with appropriate "squish" on the first layer. Always check to see if the first layer is completely stuck to the bed. If not, either manually adjust the z-axis or relevel the bed. This is a process that students should be taught early on within a 3D printing class.

Check the bed leveling process for your specific printer. Small adjustments and z-axis offset can make the difference between a great print and a failed print. If these adjustments do not fix the problem, adhesion can happen with the use of actual adhesive. Glue stick, bed spray or even hairspray can work to give a better grip. I prefer glue stick due to the availability and low cost.

Adding adhesive creates the need to clean the bed at different times. Using a razor scraper can remove residue on most build surfaces including glass, spring steel and several other plate types. Stick on beds will need either the use of a solvent, water, rubbing alcohol, etc., or be replaced at regular intervals. Remember to clean the bed surface before applying a new stick-on unit. Magnet beds come in many different forms and can be used as a print surface to quickly remove a print from the build plate. This type of plate will also require regular replacement and cleaning.

If the layers of the print are not adhering, consider raising the extrusion temperature or adding a heat bed to your printer. Heat beds can help with adhesion by creating a more welcoming environment for the filament and reduce the cooling factor after extrusion. Each type of filament will have recommendations on extrusion and bed temperature. If this does not fix the layer adhesion, your filament might have absorbed too much moisture and steam pockets are forming during extrusion and cool down not allowing the filament to properly bond. Change out the filament and properly dry in a sealed container with a desiccant.

Extrusion problems exist for several reasons, but we will cover the basic solutions, as they are the most common. Temperature of the hot end is the easiest fix. This can be increased during slicing or upped during the print in the software control. Low-temperature diagnosis happens through sound. If you hear a clunking noise during extrusion and see the filament not advancing as quickly as it should, this points to your hot end being too cold. If you are getting molten pools of plastic at varying intervals during extrusion, your hot end is too hot. If too much or too little filament extrudes, check your steps per millimeter in the firmware. Often, this number will create problems that will look like other items failing.

The mechanical problems are the hardest to diagnose and usually simple to fix. Prints not attaining full size on one axis usually points to belts being too loose or stepper motor power not being high enough. Tighten any loose belt and adjust the motor driver power on the control card. Look at the manufactures directions for adjusting these drivers. If prints are coming out backwards, typically this points to the direction of one axis reversed in either firmware, software or motor setup. Switch this within the firmware or software or if that does not work, reverse the physical connection at the card. The other way you can tell if reversed motor direction takes place is the homing sequence sends the axis away from the limit switch. Mechanical extrusion problems can present as too little or too much heat. It could merely be a clogged nozzle or infeed tube. A great solution for this is either a purchased nozzle cleaning kit or 1/16" steel wire rope taken apart to produce one strand. These single strands are a good size to feed backwards or forwards through the hot end to break up any clog. Remember to do this with either the hot end at temperature or the strand heated to the hot end temperature. In addition, the machine is constantly moving and this causes the fasteners that hold the 3D printer together to loosen and sometimes fall out. Check your printer and tighten loose fasteners every 10 to 20 hours of operation.

- Take a low-cost flat blade screwdriver and file off outside edges of the blade to create an upper feed cleaner (Figure 1.11).

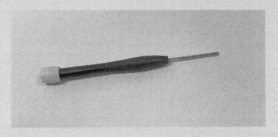

Figure 1.11 DIY feed cleaning screwdriver

- Take a strand of 1/16" aircraft cable and use as a nozzle cleaner (Figure 1.12).

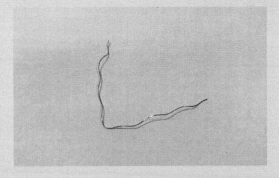

Figure 1.12 Steel wire rope cable strand

2

TEACHING AND PROBLEM SOLVING

Throughout my teaching career, I have used many articles and books to inform my idea of effective teaching. One book that has helped me is Barbara Blackburn's *Rigor is Not a Four-Letter Word* (2018). This has helped me devise my own style of creating rigor within the classroom while keeping an eye on the process. I encourage you to find the multitude of articles and books on rigor, and expand your ideas. You might find some of my suggestions later on to be more aspirational than practical, but I wish to share some thoughts with you on this matter.

With the basics of 3D printing handled in Chapter 1, it is time to consider how you can use 3D printing in your class. Most prop-making classes that I have taken only assess the final products through look, strength and application. This does not take into consideration the process or the level of growth that the student increased or the circumstances that the process progressed. These process-focused teaching moments can produce incredible results for the student and the class without putting all of the grading onto the final product. This creates a process driven by the end and not the path traveled.

I would rather hire an individual that has failed on a project and then revisited it with success than a person who has never tasted failure. The best lesson is picking yourself up and continuing to work after failing on a project. I feel that this creates the ground for learning, not only about prop design and construction, but also about the student's capability to know themselves.

First, let us explore the idea of safe failing. We also need to consider how to pave the way for engagement and acceptance by your class. What can the student gain from failing? Failure generates multiple lessons, but the most advantageous result is the ability to pick oneself up and continue working on the problem. Another advantage is the removal of grade pressure from the individual projects by allowing one or two project "do-overs" for the class, but as students will always say, "I am worried about my grades". So as one of my mentors said, take away the idea of grades. Base it on effort. That is nice and all, but it does not answer the question of effective assessment. The answer is basing assessment on process and not only on product. The product will be an artifact of the project itself and should only take up one or two assessment slots. Visualising the rubric in the final product is a common method, but the assessment criteria observation will yield more data during the process of creation. Consider using waypoints for process checking while the student works. This will also help correct any program problems and head off missteps.

The What, Where, When and Why of How

The first step is to decide when how comes into play. The question how will stop most levels of creative endeavor. It firmly puts the mind in practical modes and does not allow for continued growth of the project idea. The student will focus only on making the project work and not on growing the project fully. Do you want them to only perform the functional tasks of the lesson plan and take it no further? Yes, many of the projects in this book start out as functional only. They are the platform to get the student moving in the right direction, but it is up to you to provide the space for them to create. Get them to think and design past the surface or you will get surface understanding to create a surface item. Just as we have to develop props that help define a character and a world, the student must strive for deeper understanding from function. How should also be built up from past class skills, multiplying them through challenge and diligence. The budget of how needs to be built into the project with clear guidelines for use and what will happen if an overage takes place. How is necessary but must be placed early enough in the process to be able to have the possibility of success, but late enough to fully form the project.

The why of how is self-evident. The ultimate incarnation of the prop, viewed by the audience on stage, moves the world or character forward. This is the main idea of why. We do not create for prop's sake; we create to move the story forward. In Chapter 4, we will be looking at a candlestick phone and manipulating an existing design to fit your show or character. Have the student develop the story before the design process starts and present an analysis pre-production.

Rigor Based on Attempt and Process Instead of Product

The question for myself is this: should rigor be an impediment or bolster to the student? Should it create the environment that a student can succeed in or is competition the building blocks for a better student? Does one have to do better than another does in order to prove their academic worth? Good questions that create even more questions as you look forward. For my classes as well as the process lab in this book, I use a concept of build the student and provide space for their own exploration. Yes, this does create more work for the instructor. Minimise this by handling it with good project guidelines and rubrics. Remember, every institution will have a working definition of rigor. You might have accreditation or standards that must be met. All of these guidelines will help you create.

These rubrics will be set up based on the process with an artifact produced, the product. Consider the guidelines and rubric a contract or compact with the student. This allows the instructor to tailor the experience to each student instead of one-size fits all. What does the student see as their assets and liabilities on the project? What should be the most important learning objectives for the student? As Barbara Blackburn made popular, there are three phases of the academic process: establishing standards for students, equipping them through instruction and support and having the student achieve the standards through completion of projects, assignments, etc. (Blackburn 2018). This is an oversimplification of this work as well as a moment created in history. There will be more theories on academic rigor and process that will probably contradict my statements, but this simple process has guided me. These concepts create student buy in and ownership over the learning process. The most important objectives for myself include: clear instruction and goals; supporting the student to meet the basic standards while giving them space to explore further; and caring about process as much as product. In the first project, grades will still be the motivating factor, but after trust of the process builds, an honest assessment of self generates and a layer of academic impediment dissolves. The questions of "what did you do well?", "what needs improvement?", and "what you would change about your process?" yield amazing results after trust is built.

Games and Competition

Does competition breed perfectionism, emotional feedback and distrust? If handled incorrectly, most definitely. If competition is valued for pushing creative boundaries and generating new ideas then set rules for what the competition will entail. Never use grades as the reward of a class competition. Games also work to create challenge within the class. The game or competition within the class should not spill out to encompass other possibilities. For example, whomever wins, receives the production props design position for the next show. These selections happen because of a clear process, disseminated to create equal opportunities. Do not soften reasons for not attaining these positions but create a way forward for the student to attain their goals.

Individual vs Group

I find that individual projects should build upon themselves, and group projects should reinforce learned skills. Build in confidence by applying rubrics matched to the individual student. The onerous nature of preparing multiple rubrics for one project proves daunting but overcome it with either taking away or adding assessment items in the rubric. We will visualise this in detail in Chapter 3.

Table 2.1 Chapter 3 rubric

Area of focus	Manipulation	Alignment	Manifold	Print
4. Successful	Ability to choose and manipulate the shape to the exact desired dimensions.	All factors are aligned to the satisfaction of the design.	All areas need no repair when importing them into slicing software.	Physical print is successful and error free. When handled, it does not break.

Area of focus	Manipulation	Alignment	Manifold	Print
3. Needs focus on one area	Ability to choose and manipulate the shape but the desired dimensions are not achieved.	Alignment cannot be achieved or does not look correct.	Small gaps appear and needs repair by the slicing software.	Physical print has small errors. When handled, some small details might break.
2. Needs focus on two areas	Ability to choose and manipulate the shape but the desired dimensions are not achieved.	Alignment cannot be achieved or does not look correct.	Small gaps appear and needs repair by the slicing software.	Physical print has small errors. When handled, larger items might break off.
1. Retry with improvement plan	Manipulation of shapes are not successful.	Alignment is not attempted.	Slicing cannot be completed without errors.	Physical print has large errors. It falls apart in your hands.

For group projects, identify teams that have balance and complement each other. Divide the group to either reinforce skills handled in previous individual projects or use it as a redo for an individual student. Seek to build up the team but take time to guide with constructive feedback. No handholding is necessary, but you must meet them where they are. The overall learning outcomes need to be followed but adaption will be necessary to meet the student. Even though this generation of students have grown up with intuitive computer technology, most have never been formally, or sometimes even informally, educated with the basic skills that most of our generation took classes to learn. With this in mind, distill the movements needed without insulting the intelligence of all involved. I am reminded of a professor that took ten minutes at the beginning of a lecture to tell the present student body how the internet worked. This included how to click on a link, what links did, and how to back out of a webpage. We all thought the moment was tongue in cheek, but no, he was serious.

Meet them where they are. Most students will rapidly get to class level after minimal remedial action. Consider your population and that most have taken classes of some sort during the pandemic, and this runs the gambit of successful to complete failure. Patience and clear communication of expectations will be necessary and will need to be repeated several times.

The Process of the Process Lab

Many times, within theatre practice, we have production meetings to make sure all areas are moving forward. This mindset is only concerned with the ultimate goal, which allows the only path as start then complete. There is nothing else except "what resources do you need?", and "get it done". This "process" is good for product creation but typically does not allow for the pedagogy of theatre to have space. Even planned research and development becomes compartmentalised and the other members of the design and production team sometimes do not value the learning process. This does not really focus on the process of making theatre. If we state that our productions are laboratory extensions of our classrooms, then the process must take the lead.

The first level would be to change rubrics from product to process orientation. This can be done by having the students do a prebuild description or drawing. The intentional path of making creates a clear process to analyse. The final prop is an artifact of the creation process, not the end all be all. The second level would make space for failure and grow from it. This is accomplished with the ability to redesign a prop after unsuccessful completion with an improvement plan. The third would be to give the student ownership of the assessment creation. Ask them what areas should be made the focus of the assessment. Is it the problem-solving process of a complicated prop, or the overall finished item? This gives the students the goals very clearly. Fourth would be the student giving themselves an honest assessment of their effort. Can they offer a truthful idea of what went wrong or what challenges they faced? The final step with the process lab would be the feedback loop of self-assessment in order to make the next project more successful in all parts.

The supporting structures of the process lab would be sufficient time to create, adequate budget to create, and the tools and safety procedures to support creation. These pieces make sense to the creative mind. The harder piece is to give space for active problem solving and frustration. Walking away is okay if the student always comes back to look at the problem again. To build the trust and time to experience the process, celebrating the successes and failures equally and honestly is key. Failure is part of creation and ultimately, that is what we are here to do. Be brave in your choices and confident in the support of your team. The main enemy of the process lab is blame. Blame shifts responsibility to another person or process, and for you not to engage in your process. The ineffective methods of rigor play into the model of blame by saying that if it is less than perfect it is less than. Remember, nothing is perfect, and happenstance sometimes yields the most amazing results. We want to make the process intentional and not by luck. Have the student do this by creating a description, drawing, etc., of what they want their item to look like before they start the process. This will help them with the idea of being intentional because they have a visual goal in mind. They can then discuss when and where ideas changed in order to meet the practical work of prop making.

The process lab focuses on learning goals instead of grade attainment as the main objective. Was this the most successful prop that you could create? What would you have done better? What was controlled and what just happened? From one to five, how would you rate your overall success in attaining your learning goals? If a three, what would it take to get to a four? "Do better" is not an answer in this process. Keep the student in concrete concepts instead of abstract thoughts. Even the best artist can learn from their mistakes.

Reference

Blackburn, Barbara (2018). *Rigor is Not a Four-Letter Word*. Abingdon: Routledge.

3

LEVEL 1

Lettered Card

The lettered card, chapter organisation:

A. Design and lettering: rubric, methods, ways and means

    1. Using primitives and text to create a business card

    2. Common print problems and how to solve them

LETTERED CARD PROJECT: DUE _____

Create a 4cm by 8cm by 1mm thick rectangle that contains your preferred name with raised letters and a recessed symbol from the available symbols list. The lettering needs to be at least 6mm but no more than 8mm tall and be able to identify them with the naked eye. The recessed symbol equals 0.3mm below the upper face of the rectangle. Make sure that your item is manifold. Upload the file to your printing software, slice, print and turn into the instructor as a completed item. You will also submit your .stl file on a flash drive

---

DOI: 10.4324/9781003376484-3

to the instructor titled with your name and project 1 by: _____ date.

Conduct your assessment of the project on the following criteria:

Area of focus	Manipulation	Alignment	Manifold	Print
4. Successful	Ability to choose and manipulate shape to exact desired dimensions.	All factors are aligned to satisfaction of design.	All areas need no repair when importing them into slicing software.	Physical print is successful and error free. When handled, it does not break.
3. Needs focus on one area	Ability to choose and manipulate shape but desired dimensions are not achieved.	Alignment cannot be achieved or does not look correct.	Small gaps appear and needs repair by the slicing software.	Physical print has small errors. When handled, some small details might break.
2. Needs focus on two areas	Ability to choose and manipulate shape but desired dimensions are not achieved.	Alignment cannot be achieved or does not look correct.	Small gaps appear and needs repair by the slicing software.	Physical print has small errors. When handled, larger items might break off.
1. Retry with improvement plan	Manipulation of shapes are not successful.	Alignment is not attempted.	Slicing cannot be completed without errors.	Physical print has large errors. It falls apart in your hands.

An average score of 2.5 or above constitutes a satisfactory completion of the project. You may choose to revisit this project to increase your score and understanding.

The business or lettered card is one of the first projects that students interface with when learning 3D design processes in my classes. The project is a good base level that creates foundational processes throughout their progress in the class. These processes of manipulating basic or primitive shapes, incising or overlaying text and shapes and creating a manifold, or completely skinned, item are useful skills to use repeatedly. The rubric can be simple to start but expanded upon to create a more bountiful experience for skilled students.

First, let us start with our assets and objectives. We will be using Tinkercad, a free, web-based cad program from the Autodesk Company. I suggest using the education side of the platform and then creating a classroom for your students. This is an easy and useful workspace that you can keep track of your students' progress and offer guidance in real time if you so choose. The learning objectives for this project are as follows:

1. Students become familiar with the Tinkercad interface.
2. Students will manipulate a primitive shape.
3. Students will overlay/incise text onto the manipulated shape.
4. Students will create a singular object that produces a successful print in .stl format.

The rubric should be simple and straightforward without extraneous detail or complication to completion. At early stages, try to focus on three or four areas of growth to measure student understanding. My suggestions for growth areas are:

- Manipulation: Did the student use the software to effectively create shapes and letters?
- Alignment: Do words line up with each other and contact the card's surface?
- Manifold: Are there any holes within the skin of the item? Can they be fixed with analysis in Tinkercad or in repair in the slicing software?
- Print: Did the item actually print correctly?

The prompt either provides an open space or constrains the parameters completely. Think about where you are meeting the students and if they need guard rails or open sand box settings. Here is an example of the open prompt:

Create a rectangular shape that contains your name and a symbol of your choosing, the symbol incised, and the text overlaid on the card's surface. Remember to make it manifold, or completely skinned with no holes or open areas to the infill, and large enough for letters to look like a business card. Many times, surfaces not meeting completely will create a hole within the skin of the item. This can be fixed by re-skinning the item by increasing the polygon shells within Tinkercad or by repairing the item with the slicing software. Most cad programs that deal in 3D items will have an analysis and repair function.

This leaves a lot of room for the student to play but could seem daunting due to the sheer number of variables that exist. A constrained prompt might look like this:

Create a 4cm by 8cm by 1mm thick rectangle that contains your preferred name with raised letters and a recessed symbol from the available symbols list. The lettering needs to be at least 6mm but no more than 8mm tall and be able to identify them with the naked eye. The recessed symbol equals 0.3mm below the upper face of the rectangle. Make sure that your item is manifold and ready to print by providing it to the instructor in .stl format that is less than 1mb and labeled with your preferred name as the file name.

Saved onto your flash drive, the completed assignment is given to the instructor on _____ date.

This has all of the specifications that a student would need and gives you multiple grading points to assess. I have found that if you put the global idea first in the prompt, "Create a 4cm by 8cm by 1mm thick rectangle", and then the details, "that contains your preferred name with raised letters and a symbol that is recessed from the available symbols list". This prompt creates multiple easy to facilitate correction points. The rest of the prompt gives more detail to the entire project with parameters of the added items. First body, then limbs.

I usually add in a debrief, either singular or to the entire class, that discusses what went well, what needs improvement, what you learned and what you would change about your process for the future. Sample questions for this could include a simple "what went well?" to a complex "would increasing the number of segments in the circle create the shape you were wanting?". No matter the age of the student, this idea creates agency within the student to figure out what to do better before the instructor returns the assessment. The student should focus on accessing the learning objectives. Then, how they can leverage their personal learning style to create conditions for success.

The Project

The three items used in this project within Tinkercad: The box tool, the text tool (Figure 3.1) and any one of the symbols within the shapes library (Figure 3.2).

Manipulation of the box tool is accomplished by either dragging the reference points or by typing in the dimension desired. In this case 40mm and 80mm with 1mm for height (Figure 3.3).

The text is treated like any other shape and is resized through common word processing font size or by dragging the reference points. Apply the symbol and then turn it into a hole and recess it in the rectangle (Figure 3.4).

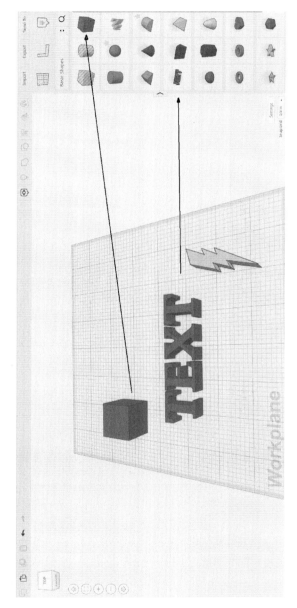

Figure 3.1 Box and text tool

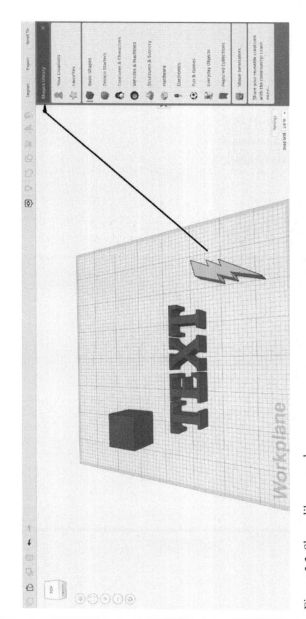

Figure 3.2 Shape library tool

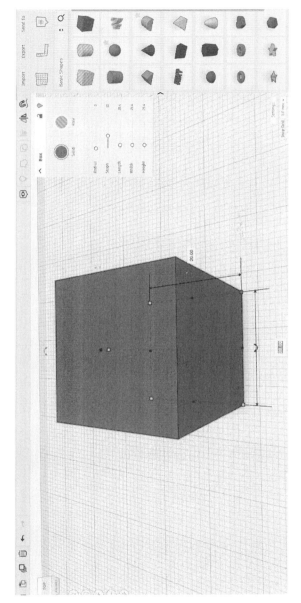

Figure 3.3 Sizing the box

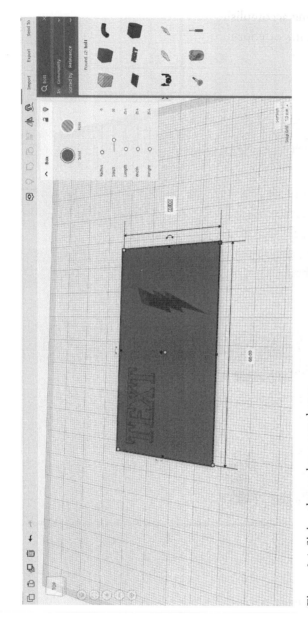

Figure 3.4 Sizing the card to your shapes

Having accomplished the procedures, it is time to create a .stl file of your item. Make sure to select no individual parts and go to export. You will then export everything in the build area as a .stl file with the title you would like (Figure 3.5).

Then it is time to slice it and prepare for printing. Refer to Chapter 1 on settings and optimisation.

The main points to observe within projects are dimensions, the text recessed on the card, i.e., touching the card surface, and the symbol being a hole and displacing area within the thickness of the card. Printing is proof, as you will end up with a lettered card or plastic spaghetti or printer blob art.

Example 1: The card will need an improvement plan in place and the student will have to revisit this project in the future. Notice the text and symbol not touching the card surface (Figure 3.6). Spaghetti for everyone!

Example 2: This card has alignment issues, and the symbol is just kissing the surface and will not appear recessed in the print. In addition, the text spans several millimeters in height and will not hold up to casual handling. The taller the text the easier it is to break off. This card would garner a solid two across the assessment board and would not meet criteria for a completed project (Figure 3.7).

Consider these items: the text being integral to the card, the symbol recessed with enough depth and alignment of

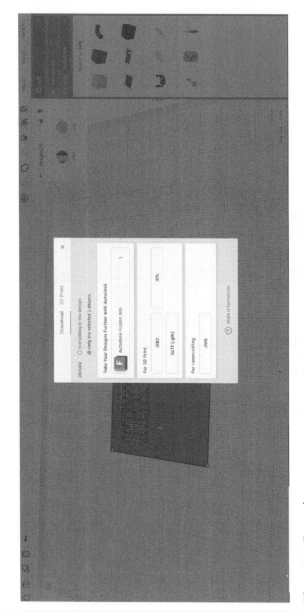

Figure 3.5 Exporting

Figure 3.6 Faulty spacing

Figure 3.7 Overlapped spacing

items to be clear and recognisable as the key focal points. Other than design problems, insufficient "squish" of the first layer resulting in corner pull up, z-axis rub on the letters causing delamination and insufficient infill density causing skin warp or droop are the most common problems in printing. Remember to check your z adjustment before critical prints, which will solve most of the adhesion issues. Please communicate these simple solutions to printing problems to the students if they are doing the printing. I would include this information within an introduction to 3D printing briefing. Keeping these ideas in mind, the student will end up with a card that is manifold and prints successfully.

How does this project work for a props class? Many items that are printed could have text or symbols recessed or raised on the surface. This adds detail and reality for the actor as well as the audience. Some details will not read from the audience, but if the detail helps the story move forward it might be worth it. The card project creates a small space for experimentation that helps familiarise the students with the basic functions of the software and printer. This could be turned into a custom stamp or wax seal project to blend with your curriculum. Use the objectives of the project to create your own. Always ask what detail will read or be beneficial to the story. This will focus the student and keep the priorities of the props creator in mind: does the props serve the story?

4

LEVEL 2

Candlestick Phone: Rubric, Methods, Ways and Means

This chapter will combine several primitive shapes on multiple planes to create a recognizable real-world item. Having learned some basic functions in Tinkercad, let us manipulate an existing prop to meet our production's needs. In this exercise, we will manipulate thing 5822811, a basic candlestick phone design that could be used in a variety of shows, displays or scenes to encourage "communication". Please download all of the files separately or in a .zip file format. "Things" are items and designs that are located on Thinigiverse, a Makerbot file-sharing site. You can locate this thing by inserting the number, 5822811, into the search box. You can then download the .stl file located in the files area of the page.

Before we start selling the manipulated items, we need to touch on the idea of the creative commons license. A creative commons license, commonly used on Thingiverse, is permission to use the designs you find for your own

DOI: 10.4324/9781003376484-4

purposes. There are usually two types: attribution and non-commercial. Attribution allows you to sell your prints but needs the original designer credited as the designer of the item. Non-commercial allows the use of the design for personal reasons but does not allow selling of said item. These two licenses can be separate but can also be combined into an attribution/non-commercial license. This is only one type of license and many more exist to protect an individual's intellectual property. Thingiverse is a place of mutual respect based on the common good so please abide by the license requirements that exist with each item.

After the download has completed, you will import the .stl file into Tinkercad. Try to import only individual files instead of the entire phone in parts. When imported, all items within the file group as one item instead of multiple items. Keep the scale in millimeters even if you are using inches and feet as your measurements. Importing using inches will equate to one millimeter equaling one foot, and your giant candlestick phone will not fit on stage due to its size.

Let us manipulate this phone to meet your needs. First, select the body of the phone that includes the upright column and round dial base. Select the entire item and use the mirror tool to reverse the placement of the micro-phone holder hook. This customises the piece for either a left- or a right-handed performer (Figure 4.1).

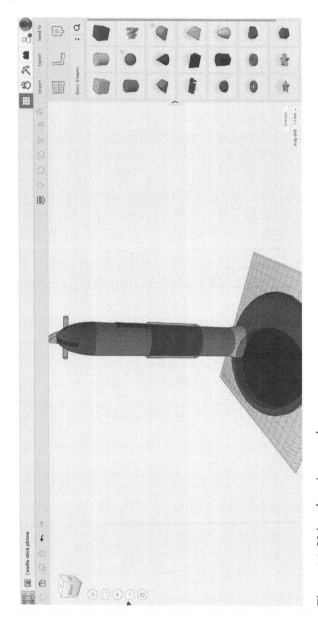

Figure 4.1 Using the mirror tool

If items were not symmetrical, this would switch the side of placement of surfaces. The next step is to shrink the piece down. Select all of the parts on your build plate, use the height dimension or z-axis, and adjust the height of all pieces down. This creates a "squishing" effect that might give you the required look for the production. You can always press the back button to restore the items to the original shape.

If you want to manipulate selected pieces as a whole, you can either select all of the shapes or use controls to resize or export the selected item and then re-import it into your workspace. This will make the item a single unit but will end the manipulation of individual pieces as a singular unit. I also maintain all of the steps of a creation process in a workspace as a point of future manipulation, but this can clutter a virtual space and might be detrimental to your overall design process. You can always import items into a clean design space to continue work (Figure 4.2).

Assembly can pose a problem if you do not plan for it. I like to check fitment in virtual before any printing is done. In Tinkercad, this can be difficult due to the simplistic nature of the program. If you want better analytical and fitment tools, use a program like Fusion 360 for final assembly. This will give you as close to a real item in the virtual world. If fitment works virtually, the next stage is to print a prototype at a lower infill rate. Think 5–10%. This will allow you to check the dimensions, fit and use while investing the least amount of material into

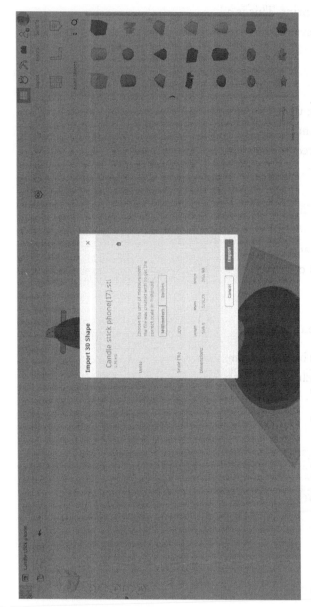

Figure 4.2 Importing a clean version of the phone

the print. This is a key step for most beginning students. Even though they printed a lettered card in the first exercise, this is their first complex print with multiple parts. A common problem for students is not landing, having the item touch the bed of the printer, when arranging them on the build plate. Several of the parts for the phone will need supports generated in order to print correctly. Make sure the student does this before slicing. Also, experiment with item arrangement on the build plate. Sometimes, only one part can be printed at a time. If printing multiple parts, arrange the lowest height items on the outside of the plate placing increasingly taller items toward the center with the tallest placed in the middle. You can also use a function called a wiping tower to help with over extrusion. Look at the documentation of your slicing software to explain the use of wiping towers and when it should be used during printing.

As before, the rubric and prompt being clear and concise with room to expand will be important for the student.

CANDLESTICK PHONE: DUE _____

Choose a play or story that this phone will occupy. Analyse the requirements and the look of the phone before design begins. This analysis will be turned in before the design process begins. Manipulate an existing prop to meet your production needs. This project is based on thing 5822811. Please change the item either in size or orientation to meet a hypothetical production's needs while maintaining fit and complete assembly. You will be using size and shape manipulation concepts from the last project as well as the mirror and multiple item selection resizing. Make sure that your item is manifold. Upload the file to your printing software, slice, print and turn into the instructor as a completed item. You will also submit your .stl file on a flash drive to the instructor titled with your name and project 2 by: _____ date.

Conduct your assessment of the project on the following criteria:

Area of focus	Analysis	Manipulation	Alignment	Arrangement	Manifold	Print
4. Successful	Story is concise and presents a clear image of the prop. Finished prop fits the story.	Ability to choose and manipulate a shape to exact the desired dimensions.	All factors are aligned to the satisfaction of the design.	Files are arranged to maximise printing success. Supports in place.	All areas need no repair when importing them into the slicing software.	Physical print is successful, fits together and is error free. Does not break during use.
3. Needs focus on one area	Story presents a rough image of the prop. Finished prop is vague but somewhat fits the story.	Ability to choose and manipulate shape but the desired dimensions are not achieved.	Alignment cannot be achieved or does not look correct.	Files could use minor arrangement help on the plate. Supports are not used correctly.	Small gaps appear and needs repair by the slicing software.	Physical print has small errors mostly fits. Feels a little fragile but does not immediately break.
2. Needs focus on two areas	Vague analysis. Prop does not fit the story.	Ability to choose and manipulate the shape but the desired dimensions are not achieved.	Alignment cannot be achieved or does not look correct.	Files need major arrangement help. Supports are not used.	Small gaps appear and needs repair by the slicing software.	Physical print has small errors and does not fit together. Pieces break off when handled.

Area of focus	Analysis	Manipulation	Alignment	Arrangement	Manifold	Print
1. Retry with improvement plan	No analysis.	Manipulation of the shapes are not successful.	Alignment is not attempted.	Files are overlapping and will not print correctly.	Slicing cannot be completed without errors.	Physical print has large errors and does not fit. Falls apart.

An average score of 2.5 or above constitutes a satisfactory completion of the project. You may choose to revisit this project to increase your score and understanding.

A debrief, either singular or with the entire class, that discusses what went well, what needs improvement, what you learned and what you would change about your process for the future is definitely needed. Consider focusing on the areas that most of the class struggle with. Try getting the students to seek peer support or seeking solutions through research. In addition, hard deadlines exist in order to put into perspective the actual time it takes to make it all fit. Complete the item with nuts, bolts and screws to speed up the assembly process. A phone cord, usually made out of tie line, finishes the phone. Painting and finishing can be required for advanced students due to the consideration for post-printing prep and priming. One of the easier ways to prime your item for painting is using an auto body priming spray paint, either a high fill or medium fill paint, or using plastidip or a similar product spray or brush on. Make sure your students are set up to do the priming and painting safely. Another fun way to interact with this project is to say the best one or two phones will live in the props storage area for future use in a production. Most students will want to keep their phones, so be prepared to print a second copy of these phones.

Additional Rigor

Manipulation of the prop by size is a simple proposition at this level. Consider having the student change the look of the phone by adjusting how many sides it has, or how round it looks. In Figure 4.3, you can see how the phone prints in its raw form. This gives the phone a scalloped look.

Figure 4.3 Printed phone

Have the students experiment with changing the feature in Figure 4.4. This could create a truly unique and character driven phone prop.

Another advanced idea is to have the character from a play chosen, the one who would own this phone, and create the prop to embody a character trait. What would it look like? What colors and materials would it be made of? The student could even explore how to make the mechanisms more refined or make the dial spring loaded, as it would be in real life. Shapes, textures and contours make a decided difference in how a character would use it.

A higher level of print would include using a variety of filaments to create the phone. How does the student's design mesh with material used? You also go for the overkill factor and make the prop functional. I would only recommend this as a high-level assignment. This would be taking the existing prop and modifying it to accept your chosen electronics. Circuitry, soldering and internal wire pathing would be needed to finish this prop. Ultimately, the question about how far to take a prop will be down to the time, budget, detail, function, safety and if it moves the story forward. Does it reveal a character trait that will be important later? How many audience members will notice subtle or not so subtle clue? Will this detail take over the creation of the prop and make the student lose sight of the overall goal of creating a prop that the actor can use? Remember, almost everything that can be classified as a prop is meant to interact with the human body in some way. Can someone actually use it?

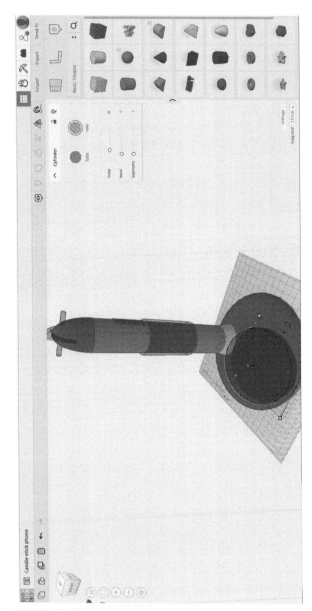

Figure 4.4 Putting more sides on the circle

Challenge Areas in Brief

1. Phone hook fitment in the phone body. This is a weak point that can break with handling. Should you reinforce it?

2. What pieces are missing from the original design that are needed to finish this prop? Could this be a team driven project?

3. How far away does the audience need to be for the prop to look like it belongs in the world? Does this distance work for the space you are using?

4. Is your conception of this phone character driven or just cool looking? What is more effective?

5. How does it feel in your hands? Would someone really use what you have created?

5

LEVEL 3

Scanning a Real-World Item

The button project, chapter organisation:

1. Scanning and achieving detail.

2. Common print problems and how to solve them.

Scanning of a non-virtual item can take place in multiple ways. Multiple photographs can be taken at multiple angles of an item (photogrammetry) and then uploaded to a program and processed into a 3D representation of an item. A touch probe put on a computer numerical control (CNC) machine which then raises and lowers the z-axis until the touch probe engages to create a 3D map of an item. Using a 3D camera to scan an item as it rotates to create a 3D file of the item. You could also design the item within a cad program by measuring the item in high detail. There are many other methods depending on your budget and time, but for this process, we will use a 3D camera with a freeware program.

If you want to use older technology that you might already have, find a Kinect camera for Xbox with the USB PC adapter. Second, search on the internet for a

DOI: 10.4324/9781003376484-5

Kinect based 3D scanner like Skanect, ReconstructMe or Scene Capture and learn the process for setup. Third, download the freeware and experiment with the software to get a good scan. Again, this is an old yet effective method for do-it-yourself scanning. You can also download any number of 3D scanning apps for smartphones, Trnio, Qlone, etc., that will use its camera or LIDAR to develop an .obj or .stl file for download. You can also buy a 3D scanner that is ready to use and have many budget options at your disposal. You will probably be spending around $500 to acquire a low-end scanner. However you decide to proceed, you will need to experiment with the device to get a decent scan.

Scanning relies on speed, consistency and detail level with most hand-held scanners. You might need to mount your scanner to a tripod or other suitably stable device to eliminate vibration error in your scan. The next question you need to answer is whether to move the scanner or move the item. You must get all angles of detail in order to have a complete scan. Do you set the item on a turntable that is spun by hand or motorised? Do you set the scanner on a moving gimbal to get all angles? Turntable of item in small spaces and moving the scanner like an airport body scanner for large items seems the best method. It all comes down to the space you have available and the item size. Controlling your distance to the object, making all angles available and consistency of movement of the scanning device are very important. General rule: if it is small and has a flat base use a turntable. If it is a statue or item in the world then walk around it.

For the button project, we will use a hand-powered lazy Susan (rotating tray) to move the button with a tripod holding the scanner (Figure 5.1).

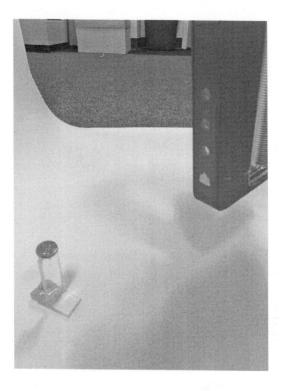

Figure 5.1 Scanner setup

Next, we need to select a button that you have one or two of, but not enough to complete your costuming project. If the button has a shank you can support it on a washer or something similar to hold the button on a consistent plane.

As before, the rubric and prompt being clear and concise with room to expand will be important for the student.

(Continued)

Area of focus	Manipulation	Scan	Manifold	Print
2. Needs focus on two areas	Ability to choose and manipulate the shape but the desired dimensions are not achieved.	The button looks close to the original button but not exact.	Small gaps appear and needs repair by the slicing software.	Physical print has small errors and does not fit.
1. Retry with improvement plan	Manipulation of shapes are not successful.	It is more "blob" than button.	Slicing cannot be completed without errors.	Physical print has large errors and does not fit.

An average score of 2.5 or above constitutes a satisfactory completion of the project. You may choose to revisit this project to increase your score and understanding.

Before your students attempt this project, confirm that your scanner can actually capture the level of detail of the possibly scanned item. If not, change the item and adjust the rubric for an attainable level of detail. Produce a clear description of the scanner rig and use and create a "common problems and solutions guide", or a CPSG document.

CPSG 3D Scanner

Problems	Solutions	Notes
Scan not accurate	Speed of turntable and distance from the item to the scanner needs to be consistent.	Slow down and be patient
Lack of detail	Adjust the resolution of the scanner within the scanner software as per the manufacturers recommendations.	

Problems	Solutions	Notes
Not scanning	Is it plugged into your computer?	Plug it in

This is just a simple example of a problem-solving guide. It will give your students a confidence boost when attempting the project. Do not single out students who are nervous or consistently cannot get a good scan. This is where your one-on-one teaching could help a student to engage and succeed rather than check out and stop trying. This is a new world for many students. They may not have had formal classes on how to use software and do not know the common process or commands to expedite their experience.

Additional Rigor

Manipulation of the prop by size and making multiple copies is the starting point. Consider having the student experiment with resolution of the printer by changing the diameter of the nozzle or layer height within the slicing program. This will increase either the detail or speed of the printing. Selecting a 0.1mm layer height will increase what you see in the detail. Have the student lower the height to 0.5mm and see if it reveals all of the detail. Look at the model in Figure 5.2. This detail will increase the time by many minutes.

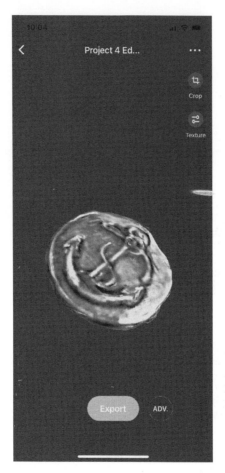

Figure 5.2 Button model

Sew the button on to the costume and see how it works. If the button has a shank, test this for shear. This is another advanced area for the student to experiment. Try different shapes and structures and then print. Remember that printers print in layers like plywood so rotating the items so the grain, or natural layer bonds, do not create a weak

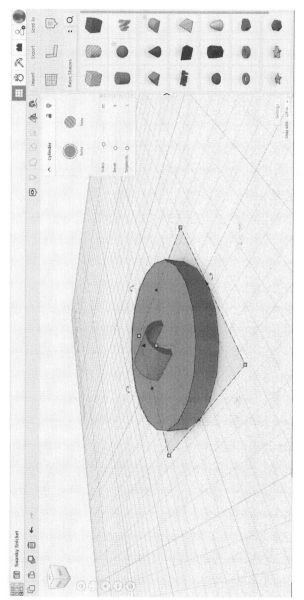

Figure 5.3 Shank design

area for the connections. Test the button on the costume and have the student record their findings. Consider Figure 5.3. Try these shapes as starting points and work your way out from that. The shank that is presented is very small and weak. Use this to show how grain in the print can create break lines and weak points.

Another advanced idea is the painting or decoration of the button. What type of paint and primer is needed? Have the students do research into coatings that would hold up under costume use. Do these give the button a real look? Do you need the button to look exactly like the original? Are they distressed or altered to embody the character?

Achieve a higher-level finish by "baking" the button in a toaster oven at 110–160°C. This would meld the seams together to form a cohesive whole. This type of annealing can cause shrinkage of the print and is usually only done for certain types of PLA filament. Does this also strengthen the button?

Challenge Areas in Brief

1. The shank of the button might break. This is a weak point that can break with handling. Should you reinforce it?

2. What sizes of button are missing from the original design that are needed to finish this costume? Do you need to make smaller or larger buttons? Could this be a team driven project?

3. How far away does the audience need to be for the button to look like it belongs in the world? Does this distance work for the space you are using?

4. How does it feel in your hands? Do they actually work for their intended purpose? Always evaluate for time and budget constraints.

6

LEVEL 4

Makeup Prosthesis: Rubric, Methods, Ways and Means

The makeup prosthesis project, chapter organisation:

A. Flexible materials

1. Safety and concerns about gluing things to your face.

2. Sculpting, slicing and making manifold.

3. Common print problems and how to solve them.

Let us start gluing things to our faces. This is a line that you were probably not expecting. We will be using flexible material and various stage makeup glues to create a reusable prosthesis, in this case a nose and horns, which adheres to our face and can reproduce copies with a click of a button. This is not typically within the props person's production area but is a logical progression of 3D printing for the stage. Be mindful of depictions that are racist, sexist, etc. Guide your students to conduct themselves with sensitivity to other cultures, races, etc. Exaggeration of facial features or creating stereotypical depictions of anatomy can be insensitive and destructive to your community.

DOI: 10.4324/9781003376484-6

Flexible filament is a strange beast. Temperature is a finicky thing when dealing with flexible material. It will either flow perfectly, not extrude or bunch up in the extruder feed drive. This is a matter of 10°C so testing is necessary. I typically use one of my older printers, a prusa i3 clone, to make flexible prints. It is easier to feed the filament and easier to see any mess-ups during the print. It is also a direct drive extruder that has variable filament tension settings. The flexible filament we will use in this project is thermoplastic polyurethane (TPU), it is a non-reactive material that, according to the safety data sheet (SDS), contains no chemicals that would harm you from skin contact. You must always check to see if a particular filament is safe for use on skin. Please continue to check this data as new tests confirm different findings for newer products. Remember a filament from one company might not have the same SDS as another for the same material.

As with Chapter 5, we will be using a 3D scanner to produce a starting point for the creation of our nose and horns. We will also use a program from Autodesk called Meshmixer to sculpt our prosthesis. You might notice that I am using older programs to complete these projects. Accessibility in price and low-learning curve time are guiding principles in this book. Fusion 360, Blender or other software can be used to accomplish the same results.

As will all scanning, resolution and patience are key to a good scan. Carefully use the scanner to scan your nose and forehead in order to begin work. After the scan is finished and the model generated, import the scan into Meshmixer. I usually will use the sculpt brush in order to manipulate the scan. Remember to leave a space for the nose to occupy. In this case, I am creating a nosepiece for a tree creature with branch horns to match. Experiment with the different features of Meshmixer to get a feel for it (Figure 6.1). I have found sculpt is the easiest to translate to the skills learned while obtaining a theatre degree.

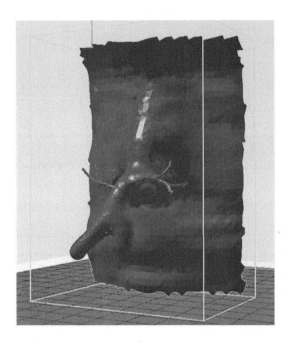

Figure 6.1 Nose design in Meshmixer

Once you are satisfied with your nose, export it as a .stl file and move it into your printing interface. Think about using the flexible material on the lower temperature side of its working temperature corridor to start with. If you are finding that no matter what temperature you are printing at that the filament bunches up in the extruder feed, you need to add a spool break. This gives resistance to the filament so it does not have time to heat to an almost melting point before entering the hot end. I usually do this by using a spring clamp or clip to ride against the spool to add a slight amount of friction. There are many other ways of accomplishing this, but I had spring clamps in my lab that fit the bill (Figure 6.2).

Figure 6.2 Spool break

After cleanup is completed, try the piece on for fit. If you need to change the shape for better fit, use a heat gun to move the plastic into a better shape. After this, get gluing. I usually use spirit gum, but there are any number of products that will accomplish this process. Baby wipes and specific glue remover for the product that you use will be necessary to remove the prothesis. If you feel any discomfort or skin irritation, either adjust or stop wearing the prosthesis and use a different filament, flexible PLA or the like, and try again.

Some common problems will likely come up during printing. Other than initial feed issues, resulting in lack of print, lesser feed issues will cause thin shells and lack of infill. These could result in slumped or incomplete prints. Counteract these problems by adjusting the tension on the extruder feed drive or by using a fan to cool the filament as it enters the hot end. Lines can also form, marring the print slightly. Overcome them with the use of makeup after applying or acrylic primer before applying to one's face (Figures 6.3 and 6.4).

Figure 6.3 Freshly printed nose, notice the lines of the layers

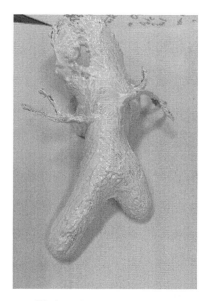

Figure 6.4 Nose filled to eliminate printing texture

Create masks in the same way as the nose prosthesis. Scanning of the face to create a surface to build the mask is commonplace. They will fit well, and will only need minimal adjustment for each actor. The use of mole-skin or other blister reducing material will help with sweat and comfort. The masks shown in Figure 6.5 are Commedia masks printed in a flexible PLA material. They are robust, hard to damage and will return to shape even when crushed. These masks only took four hours to print. They print rapidly to accommodate for changes in the production. This project should only be shared with more advanced students as a supplementary project. Since it is dealing with fit and comfort level of the actor, it is hard to adjudicate in a classroom setting as conditions are always in flux.

Figure 6.5 3D-printed Commedia masks with additional painting by Professor Casey Watkins

That being said, let us set up the project rubric with the challenges in mind. Remember that failure is a great tool to teach resiliently and a positive attitude. A level of caring with subtle pushing is the teaching method that will probably work the best.

MAKEUP PROSTHESIS OR MASK PROJECT:
DUE _____

Please scan and create a .stl file of your face that can be used to build your mask or prosthesis. You will be using a 3D scanning rig to generate the model and then size and shape manipulation concepts in order to make your face-changing item. You will also use the sculpting tools in Meshmixer to help in the creation. Make sure that your item is manifold and ready to print by providing it to the instructor in .stl format that is less than 1mb and labeled with your preferred name as the file name.

Saved onto your flash drive, the completed assignment sent to the instructor by: _____ date.

Conduct your assessment of the project on the following criteria:

Area of focus	Manipulation	Scan	Manifold	Print
4. Successful	Ability to choose and manipulate the shape to the exact desired dimensions.	Detailed scan that translates to creation of a mask or prosthesis.	All areas need no repair when importing them into slicing software.	Physical print is successful, fits together, and is error free.
3. Needs focus on one area	Ability to choose and manipulate the shape but the desired dimensions are not achieved.	Most of the detail of the item is apparent.	Small gaps appear and needs repair by the slicing software.	Physical print has small errors but mostly fits.
2. Needs focus on two areas	Ability to choose and manipulate the shape but the desired dimensions are not achieved.	The face looks close to the original but not exact which creates problems with fit. etc.	Large gaps appear and needs repair by the slicing software.	Physical print has small errors and does not fit.
1. Retry with improvement plan	Manipulation of the shapes are not successful.	It is more of a "blob" than anything else.	Slicing cannot be completed without errors.	Physical print has large errors and does not fit.

(Continued)

An average score of 2.5 or above constitutes a satisfactory completion of the project. You may choose to revisit this project to increase your score and understanding.

Additional Rigor

Manipulation of the prosthesis in Meshmixer is the basic level. Consider having the student experiment with thinning the edges of the prosthesis to aid in gluing. This should give better "wings" to attach it to the face. Are there any adhesion problems? Do the items stick with movement or sweat? Can this be addressed in the design or printing of the prosthesis? In Figure 6.6, some additional add-ons are used to help in adhesion.

Another area of expansion could be breathing. The creation of breathing holes is necessary, but not all breathing holes are the same. Does the prosthesis change the sound of the actor's voice? Can a design or printer change, correct or augment this effect?

Another advanced idea is the painting or decoration of the prosthesis button. What type of paint and primer is needed? Do these fumes cause irritation after it is dry? How does it blend into the skin and does makeup adhere to it? Do you need to add texture to the prosthesis in order for makeup to stick? Are they distressed or altered to embody the character?

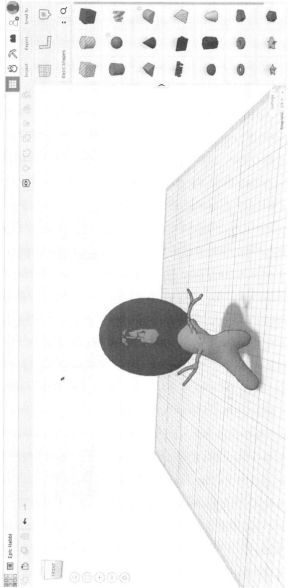

Figure 6.6 Nose with thin layers of material to increase adhesion

Challenge Areas in Brief

1. Remember breathing holes. Do you need to decrease the infill to reduce the weight of the prosthesis?

2. Do the horns and nose match? How many people would be wearing these on stage? Could this be a team driven project?

3. How far away does the audience need to be for the prosthesis to look like it belongs in the world? Does this distance work for the space you are using?

4. How does it feel on your face? Do they actually work for their intended purpose? Always evaluate for time and budget constraints.

Remember to use any guides that will help the student to become independent within the creation of this project. Even though this is a difficult project, focus on building confidence in your students. Your ultimate goal is for the student to attempt these projects and come to you for specific help. Build the scaffold of support with each project so that students can start to support themselves.

7

MOVING PARTS

Motion and control of said motion is sometimes necessary within props construction. Typically, this will be rotary motion turned into linear motion, but for this project, we will move through simple to the more complex. A basic skill of simple soldering will be necessary to complete this project. You will also need some DC motors, which can be salvaged from toys or old printers, or you can purchase type 130 miniature DC motors. These are not powerful motors but will do nicely for this project. Let us start by using the gear creator located in Tinkercad (Figure 7.1). There are several methods to create gears so please use which ever one you would like.

DOI: 10.4324/9781003376484-7

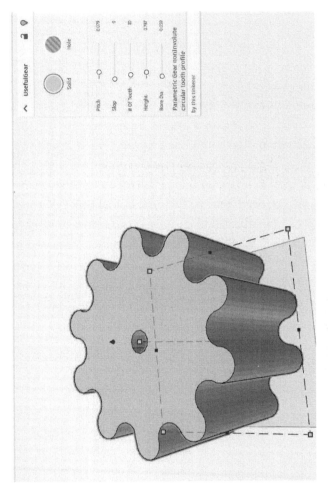

Figure 7.1 Tinkercad gear creator

Use this feature to create two gears, one large and one small. I like working on a piece of construction foam or corkboard in order to "pin" moving components together in a drive train to make sure the intended movement happens. Remember, that you can find many proven drive trains online and printed out, so no experimentation is necessary. For this project, we will focus on making our own in order to learn the process of creating motion (Figure 7.2).

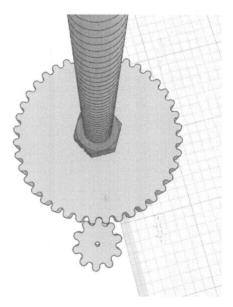

Figure 7.2 Drive train box

Ensure that both gears have a central hole that will fit your chosen motor, and use small nails or pins to hold the other gear. You might need to bore out the holes for a more exact fit. Now wire your motor to a 1.5–3 volt battery and see what happens. The next process, if this drive train meets your needs, is to create a gearbox that holds the mechanism, and lubricate with a type of grease such as white lithium.

The next step is to create a linear actuator. This translates the rotary motion of the motor into a reciprocating motion (Figure 7.3).

Figure 7.3 Linear actuator assembly

You will use the gear dive from the first experiment. Attach a threaded rod to the large gear and use a tube with a channel cut out to hold the nut in position. When the threaded rod spins, the nut moves up and down its length. Rotary to linear. This can actuate many things including wings and lifts.

LINEAR ACTUATOR PROJECT: DUE _____

Please use the Tinkercad gear generator to create the drive portion of the linear actuator. Then create a drive train and housing in Tinkercad and test your actuator. Make sure that your item is manifold and ready to print by providing it to the instructor in .stl format that is less than 1mb and labeled with your preferred name as the file name.

Saved onto your flash drive, the completed assignment sent to the instructor by: _____ date.

Conduct your assessment of the project on the following criteria:

Area of focus	Manipulation	Function of actuator by itself	Usage	Print
4. Successful	Ability to choose and manipulate the shape to the exact desired dimensions.	Mechanism works with few problems.	Mechanism opens the box successfully. Evidence of successful problem solving observed.	Physical print is successful, it fits together, and is error free.

(*Continued*)

(Continued)

Area of focus	Manipulation	Function of actuator by itself	Usage	Print
3. Needs focus on one area	Ability to choose and manipulate the shape but the desired dimensions are not achieved.	Some problems in function but they can be overcome with a small re-design.	Mechanism opens the box with some problems. Evidence of some problem solving observed.	Physical print has small errors but mostly fits.
2. Needs focus on two areas	Ability to choose and manipulate the shape but the desired dimensions are not achieved.	Some problems in function but they can be overcome with a large re-design.	Mechanism opens the box with many problems. Evidence of little problem solving observed.	Physical print has small errors and does not fit.
1. Retry with improvement plan	Manipulation of the shapes are not successful.	Does not function.	Does not move the box lid.	Physical print has large errors and does not fit.

An average score of 2.5 or above constitutes a satisfactory completion of the project. You may choose to revisit this project to increase your score and understanding.

Problems Solving with Electrical Connections

Your students might not be versed in electrical connections or how batteries work. Remember the simple analogy of the negative flowing to the positive. You could clue this into human emotions by saying: The negative wants to be happy, so it will surround itself with positive by going over to the positive side. You could also make a Star Wars reference about the light side and the dark side, etc., if that works better.

Break it down for the student. Is there power? Is the battery hooked up correctly and does it have a charge? Is the power moving from negative to positive with continuous wires and devices? Does the device work? Teach them how to test their components before installing them into their props. A multimeter or other testing device would be helpful in this experiment. Set the multimeter to continuity to see if the circuit is complete. You might need to teach the students on multimeter or test lamp function to make this part of the lesson work. Finally, are the connections solid and conductive? Many times, students will cover the connections in tape and not worry about them again. Try to get them to use solder, wire-nuts or other connections that do not require tape. This basic trouble-shooting guide will not only help your students but help you to problem solve with them.

Additional Rigor

To create a more advanced project, have the students research DC motor control and apply that research into controlling their linear actuator. The basic idea is that reversing polarity, switching positive and negative leads, will cause the motor to go in forward or reverse. Add more gears or the size of gears to decrease the speed of the threaded rod and increase the lift ability of the actuator. Conversely, switching the gear ratio to large on the motor and small on the threaded rod will increase travel speed but decrease lift potential.

To add further advanced projects consider having them create a rack and pinion, worm gear or even a selectable speed transmission. On the electrical side, you could introduce variable speed or even Arduino controlled servos and stepper motors (Figure 7.4). These projects must do something other than exist. Use them to remotely open a mailbox, move the eyes in a creepy portrait or create a plant that magically grows and produces a flower.

Figure 7.4 Arduino Nano controller

This might be too much as it will need to branch the project beyond 3D printing and into coding and electrical theory. Remember to keep the rubric contextual and clear with objectives that can be obtained.

Challenge Areas in Brief

1. Gear meshing and precision of prints are a challenge in this project.

2. Does the gear ratio you selected work for your project? Do you need to gear down or up?

3. What kind of projects would this work in actuating? The actuator is a tool. You must use it to solve a props problem.

8

LARGE PRINTS USING SMALL PRINTERS

In this chapter, we will briefly discuss the process of cutting large pieces of a print into smaller pieces to go on the build plate of an eight-inch cube printer. As we have discussed in previous chapters, the eight-by-eight-by-eight printer is a ubiquitous size within the premade or kit printer world. For larger prints, you can either cut the item up into smaller sizes or purchase a large format printer.

The first question will be what type of adhesion method will be used to "glue" the parts together. Physical add-ons could be places for screws or bolts to align and hold parts (Figure 8.1).

This can be time consuming and possibly dangerous if the print fractures due to screws or other sharp metal pokey things.

The second method would be using glue to adhere the pieces together. This can be a stronger bond when connecting pieces if the correct conditions are present. Polyurethane glue is a great idea. It has strong adhesion to PLA prints and can be sanded once cured. The downside is the foaming nature of the glue due to moisture

DOI: 10.4324/9781003376484-8

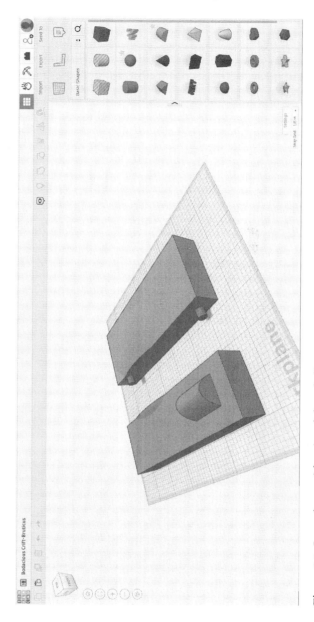

Figure 8.1 Large item with printed dowel alignment

in the air. Follow the guidelines for the adhesive and try to find a low-foam product. Super glue can also work. This needs a completely flat surface to bond to in order for adequate connection. If the surface is not perfect, use a thick "CA" glue from a RC plane store to make sure bonding is successful. Do not use a high-temperature hot glue or construction adhesive, as either the temperature or the solvents will deform the print.

The last method that I have used is automotive double stick tape. This causes the items to be slightly larger due to the tape layer but can provide an excellent bond (Figure 8.2).

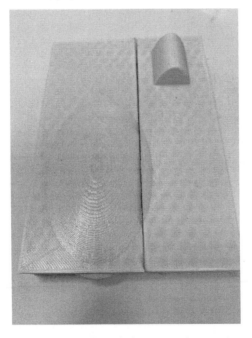

Figure 8.2 Using double sided automotive tape to adhere pieces together

The finishing process completes with either sanding the seams or filling the seams. I have used acrylic caulk, low-temperature hot glue and hot forced air to blend the seams. Hot forced air can be a paint stripping heat gun to a solder forced air rework tool. Many other methods can be researched and used. Encourage your students to discover safe and quick methods to accomplish their goals.

The final phase will be priming. Again, you can use plastidip, automotive primer, acrylic paint and even gesso to prime the plastic. Remember, you will need to test the item for damage or rub off if handled extensively. A sealing coat might be necessary, but always plan for a repaint time schedule if needed. Finish painting can then be taken on after the multiple prints are primed (Figure 8.3).

For my production shop, statues and busts have been the most common multi-part print. With time and patience, almost anything can be built to any size. Many times, I will add dowels or pegs to help align the items. Think mortise and tenon using integrated plastic dowels and holes. It is essential that you cut at logical breaks, not where the seam will be obvious. You can translate that into cutting along the outside of the face, not directly in the middle of the face (Figure 8.4).

Plan your cuts to give the best visual product instead of using a grid pattern or easiest cut line. Also, plan a flat surface to be the base of the print and build up from there. You can also inset reference letters, just like the lettered card project, to help with larger prints. This will

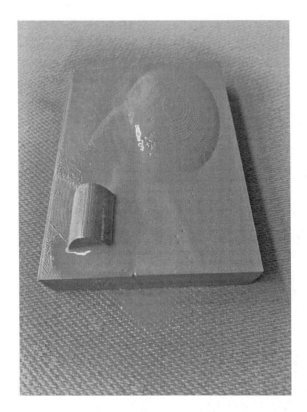

Figure 8.3 Primed and painted

not only give a visual to assembly, but also ensure that the pieces are adhered together in order. Start from the bottom or start at the highest detail area and work your way out.

Clamping or taping will be necessary to hold the pieces in place. I use a dedicated worktable to do large print assembly. Vibration of the work area is one of the greatest

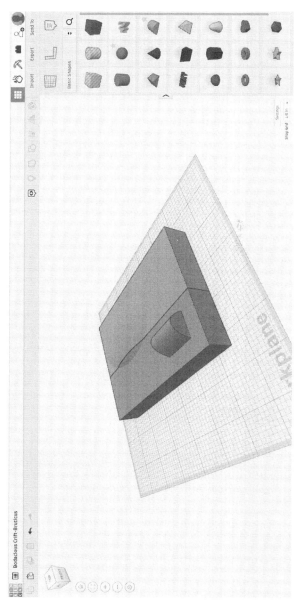

Figure 8.4 Cutline over face of item

challenges. Be mindful and patient and your assembly will go well.

Now, let us get into a project. This will be a group project with at least two students working together. Each student will work to print the cut pieces and then combine them with their partner's printed pieces. This is a good way to introduce working together with just a few assessment points at stake. Final fit within the rubric will assess the final product, but process steps of problem solving will be the real judge of group effectiveness.

LARGE PRINT GROUP PROJECT: DUE _____

Please use the Tinkercad featured collections tab to choose an item to divide up. Have each student print at least one part. Then combine the prints into the selected item and have them prime and paint the assembled item. Make sure that your item is manifold and print and submit the finished product along with all files to the instructor in .stl format that is less than 1mb and labeled with your preferred name as the file name. Saved onto your flash drive, the completed assignment sent to the instructor by: _____ date.

Conduct your assessment of the project on the following criteria:

Area of focus	Manipulation	Team process	Manifold	Print	Final fit	Finish
4. Successful	Ability to choose and manipulate the shape to the exact desired dimensions.	Problem-solving process is discussed and agreed upon. Conflicts are resolved.	All areas need no repair when importing them into slicing software.	Physical print is successful, it fits together, and is error free.	All parts make a whole.	Painted and looks like a single item.
3. Needs focus on one area	Ability to choose and manipulate the desired dimensions are not achieved.	Problem-solving plan is rudimentary and not used. Conflicts are somewhat resolved.	Small gaps appear and needs repair by the slicing software.	Physical print has small errors but mostly fits.	Some problems with final fit but are resolved with physical manipulation.	Painted but small flaws apparent.
2. Needs focus on two areas	Ability to choose and manipulate the shape but the desired dimensions are not achieved.	Problem solving is done on an individual level and conflict results from poor communication.	Small gaps appear and needs repair by the slicing software.	Physical print has small errors and does not fit.	Gaps and spaces are present but can be filled.	Painted with large flaws.
1. Retry with improvement plan	Manipulation of the shapes are not successful.	This is a giant mess.	Slicing cannot be completed without errors.	Physical print has large errors and does not fit.	Not fitting together.	No finish at all.

An average score of 2.5 or above constitutes a satisfactory completion of the project. You may choose to revisit this project to increase your score and understanding.

In Tinkercad, find an item in the featured collections tab. For my project, I choose a Triceratops skull (Figure 8.5).

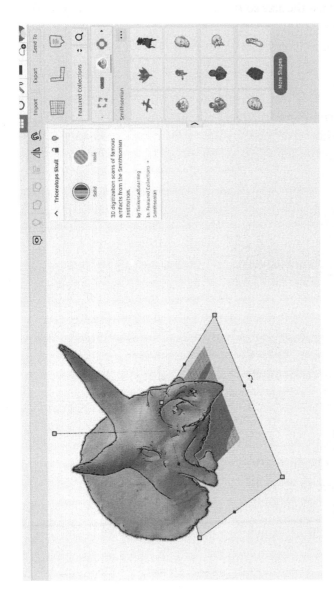

Figure 8.5 Tricerotops skull

Adjust the size so that it is larger than your printer build size. It does not have to be much larger but try for at least three separate pieces to print. Remind the student to consider the adhesion process that they wish to use and whether they need tabs, pins or to add space for tape.

Next, use the box tool on the hole setting to start to divide the item. The Triceratops skull will be divided up into four pieces. Remember to save each quarter as a separate .stl file (Figures 8.6 and 8.7).

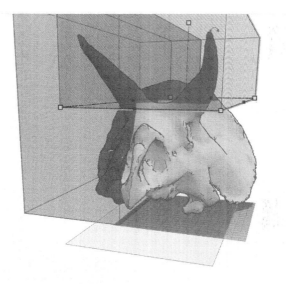

Figure 8.6 Cut lines

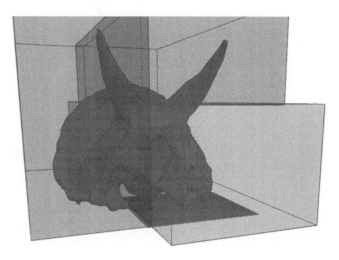

Figure 8.7 Cut complete

Have the team then combine their individual cuts in virtual before printing (Figure 8.8).

Figure 8.8 Four parts

After assembling the final print, have the team prime and paint the item. This will be the finish product but a post turn in reflection could be useful to determine team process.

Create either a journal or reflective prompt so that each team member can give feedback on the process. This can be done in many ways in order to get honest feedback about their process. I prefer either written journal entry or in-person singular and team debrief.

9

ADDITIONAL PROJECTS

Science Fiction Communicator

In this chapter we expand the scope of additional projects by delving into different products and genres.

The science fiction communicator project is a staple of one of my earlier property making classes. It combines the need for the audience to recognise that the device is for communication, along with what communication devices could be from the future or an alternate universe. Have the student choose to create their own universe or expand upon their favorite existing one. This should not be a replication of an existing device but a new creation or expansion of an existing one. There are two advisable methods to go about this project.

The first method is to use the 3D printer to create various parts so the student can pick and choose their creation. I would recommend printing a variety of cases or bodies, buttons, microphone devices and screen-like items. Have the students use foam clay to alter the shape of the bodies if one does not suit their needs. They can also scavenge parts from old cellphones, etc., that are available for free. The goal is to create a device which looks like a future

DOI: 10.4324/9781003376484-9

item that the audience will understand is for communication. Consider any of the major science fiction franchises. They use devices that are just slightly above the technology that we have today (Figure 9.1).

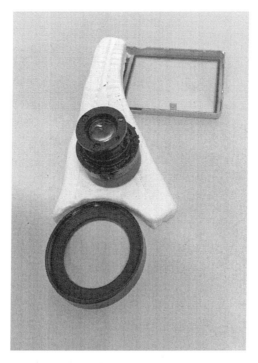

Figure 9.1 Parts communicator

The second method is to create a device in virtual and then print it as a whole. This method challenges the virtual design ability of the student. Printing the whole makes the device seem manufactured and polished. The same rules apply with the audience and its suspension of disbelief. It must seem like it would be used as a communication device and be believable. Taking this device to extremes will make the audience doubt that it works. Basing it in a slightly more advanced reality is the best path (Figure 9.2).

Figure 9.2 3D printed communicator

Assessment areas for this project: believability of the device, fit in the world chosen and the look of functionality. This project treats the 3D printer as a mere tool instead of the primary learning objective. This project is a great way to engage student interest and creativity. If science fiction is not to your students' taste, try steam or diesel punk devices.

Museum Replica

As the heading implies, the student will create a piece that could be used in a recreation or museum display. This is a niche variety of prop making but can be a real opportunity for employment in the future. Again, the main focus of this project will be in the process of creating the overall item and not 3D printing alone.

Selecting the recreation period and item might be a difficult task for the students. Consider restricting the possibilities by determining the period for the students. If the item is available for analysis, you could turn this into a 3D scanning project. Are there pieces missing that need to be recreated? Using a scan to create the base and then building the missing items in virtual could stretch a student's ability to create in Tinkercad, etc.

Another area of focus would be the recreation of color, texture and patina. Combining the creation of shape with the printer and traditional painting techniques would challenge the student with the new and old (Figure 9.3).

Figure 9.3 Ink well and quill stand

Another project that has a smaller scope within this branch of prop making would be recreating handles and hardware for antique furniture. Careful measuring of the existing handles will be challenging for some students. Prepare to guide them in precision usage of rulers, tape measures, etc. In Figure 9.4, we see a mock-up of a handle for an antique walnut teacher's desk.

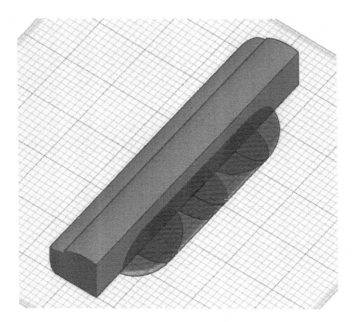

Figure 9.4 Desk handle

Painting will be a major component to complete this project. Make sure to have primers, acrylic paint and some type of clear paint sealer to apply over the item at a minimum. Possible assessment areas could be structure, replication accuracy and finish. The structure would consider the 3D printed item before paint, replication accuracy considers measurements and scale of the item and finish would take the finished product representing the actual item. Is it believable and does it look like the original item? That is the overarching goal of the museum replica project.

Jewelry

This is a project that takes some of the same principles as the button project but can engage students' interest due to the creativity that can be applied to an item. The obvious choices are rings and pendants. The key to guiding the students is to focus on detail that can be discerned by an audience instead of the actor. Strength of materials will also play a part in choosing filament. Due to PLA's brittle qualities, PETG or ABS might be a better choice. Remember to adjust your temperature settings to accommodate the filament chosen. This project would be a good team exercise. It is better to have the students design and print and item that another student wears instead of a wearable for themselves. On a practical level, most props artisans will be creating items for actors and not themselves. On a fitment level, you can deal with ill-fitting materials if you are making them for yourself. It is better to experience the interaction of fittings and changes due to this process before you are thrown into a full props design with no time available to make the item multiple times to get the fit right. Figure 9.5 shows an example of what is possible with a simple design.

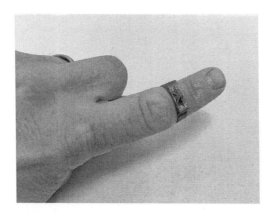

Figure 9.5 Ring

More complex details can be added to compliment the choices made for character and visual aesthetic. For this project, I would recommend having access to different jewelry catalogs like Hoover and Strong or Rio Grande. These catalogs have parts and pieces displayed that could be combined to create any number of jewelry combinations. You can also get into period jewelry with books like *An Illustrated Dictionary of Jewelry* by Harold Newman (1994) or *Antique Jewelry: Its Manufacture, Materials and Design* by Duncan James (2008), among others. There are literally thousands of resource images of period jewelry.

Think about creating a project that either replicates a famous piece of jewelry like the Hope Diamond setting or a piece of Tutankhamun's treasure. It could also focus on reinforcing a character through design choices. The jewelry project can have multiple pedagogical focus points. These can include character suitability, age, class, wealth, status and function, and detail. The main focus area must be reinforced in the assessment areas: character suitability will give information to the audience that is not spoken, and status of the character would indicate position within the society that they occupy. This project can get out of control easily so I would recommend being very clear in the prompt and assessment points.

This might be a time to experiment with different nozzle sizes. Some slicing programs may not support changing nozzle sizes. This might be done in firmware settings. I would recommend checking your firmware manual or website to check on nozzle support. You can also check the RepRap wikipedia entry. It is a great source for new and older configurations, firmware varieties and hardware setup. There are also some printers specifically for jewelry printing. These printers vary in price and technology. Companies like Soldiscape, Etec and Formlabs have several printers that will expertly print wax-based jewelry models that can be cast in a refractory material, the wax melted out and then the void filled with metal. There are also metal jetting and laser sintering printing that use different technology. If you want to learn more about metal printing, please take a look at manufacturers' websites or in books like *The 3D Printing Handbook: Technologies, Design and Applications* by Ben Redwood, Filemon Schöffer and Brian Garret (2017).

Tools

It might be counter intuitive to think about 3D printing tools, but it is possible and useful to do so. I like to divide up tools into useful categories that are needed in a typical properties shop. Clamps or holding tools, measuring aids and centering jigs. These are simple devices that would make practical projects for your students. There are many other tools and designs out there, but I chose these due to non-impact related use and ease of printing.

Clamps

Clamps are one of the easiest tools to make with a 3D printer. Glue together your pieces and place the 3D printer on top. Instant clamp! If you actually want to print the clamp pieces in plastic, it will probably fit into the narrative of this undertaking better. Consider a standard spring clamp. It has two arms that function as spreader handles and a spring or band to push or pull them closed. They create compression between the two arms. There are many designs of spring clamps on Yeggi and Thingiverse, but you might want to assign a team to create their own. You can either have them design a basic clamp or one that is special purpose to solve a unique situation. I would recommend using rubber bands for the compression engine due to their inexpensive nature and their availability in most settings. Remember, most spring clamps have two arms, a pivot connecting the two arms and a compression engine. Figure 9.6 illustrates this simple design.

Figure 9.6 Spring clamp

For additional ideas, ask them what a spring clamp can do. Have them consider other uses by asking them if clamps could be used in hair care, medicine, paleontology, etc. Use hair ties or surgical tubing as the compression engine, or completely change it around by having them create a spreader clamp, or a clamp that moves items apart.

Measuring Aids

This idea is not just about printing rulers but creating spacing aids that will make doing a repetitive task faster and easier. Most majoring aids will help you center something or create drill points for mounting handles. Figure 9.7 is a good example of a measuring aid.

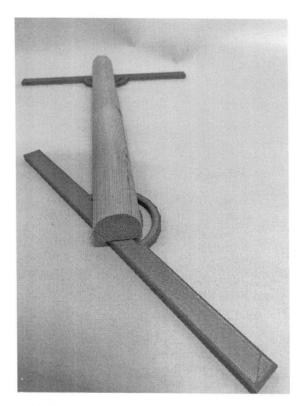

Figure 9.7 Stair tread measuring device

This device allows for exactly marking the size and shape of a stair tread onto a new piece of wood so that it can be cut and replace the damaged tread. It is two bolts, two printed pieces and a small piece of wood. Another device that could be made is a hem line measuring stand (Figure 9.8).

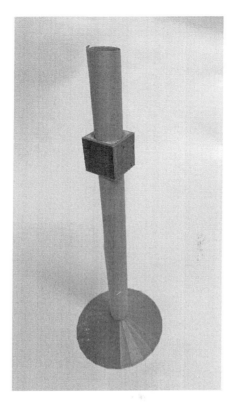

Figure 9.8 Hem stand

This device is invaluable within the costume shop for hems and can be used in the prop department for hemming all types of curtains while in place. The measuring aid project could be an invention project to find a solution to a common problem within the shop. Your students could create a device that truly makes a task easier for a show. Have them think about repetitive tasks that they perform and if a measuring aid could alleviate some of the prep or process time during the task.

Centering Jigs

As the name implies, centering jigs help you to find the center of an item. Typically, the end of a board or face of a drawer. These can be very simple or more complex depending on the function you wish to complete. Most of the time, I use centering jigs to help determine the exact center of the end of a board so that it can be turned on a lathe (Figure 9.9).

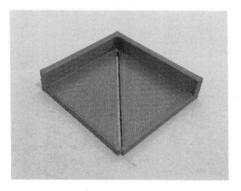

Figure 9.9 Center line tool

Instead of lining up the exact edges and then drawing a straight line between them, a simple device that you could place on a corner could draw the line every time. This is a very basic idea and device. Start at this point and move out to more complex. Ask your students what things need to be centered. Would a specialised device help in making this faster? Would it take longer to use this device than using a straight edge or square? Developing an invention just because it would be cool usually does not meet the criteria or the need to create a labor-saving device.

These additional project ideas should spark more projects that will benefit you or your students within their given circumstances. Consider these questions: what is possible to develop within your timeframe? Does the filament or material at your disposal lend itself to the device creation? Do you have enough printers or printer cycles to complete every group's items? These questions will help determine whether you should assign a project. Make sure that the projects are challenging but are also possible to complete within a given time. Let me give you an example. A well-trained technical director that I worked with insisted that platform legs needed a 25-degree bevel cut on either short side.

The idea behind this was solid. Decrease the outer edge of the leg so that it is not affected as much by uneven surfaces. As the students began cutting these legs, I started timing how long it took. The initial length cut took about 30 seconds for a trained but unskilled student to cut. Each 25-degree cut took an additional 35 seconds per cut making the overall time of cutting a leg, leaving out setup and prep time, to 1 minute 40 seconds. That might not seem like much but cutting the 450 legs for this show added roughly an additional nine hours in cutting. If you want to work this out to its logical end by cutting those degrees due to a preference of one person it added $500 in labor, $60 in electrical cost and about $150 in replacement blades and maintenance. A little over $700 for a preference. Make sure the device is needed and will save time, money or create a safer work environment.

References

James, Duncan (2008). *Antique Jewelry: Its Manufacture, Materials and Design.* London: Shire Publications.

Newman, Harold (1994). *An Illustrated Dictionary of Jewelry.* London: Thames and Hudson.

Redwood, Ben, Schöffer, Filemon and Garret, Brian (2017). *The 3D Printing Handbook: Technologies, Design and Applications.* Amsterdam: 3D Hubs.

10

SUM UP

Throughout this book, we have been exploring basic concepts and methods to quickly teach and evaluate 3D printing. This is not an exhaustive tome on 3D printing or props making. These projects are starting points. If you and your students are new to 3D printing, I would assert that these projects will take more than a semester-long class. If you have a passing familiarity with 3D printing, the projects in Chapters 3–8 can be done in one semester. This will depend on how many printers are at your disposal and what type of filament you use. Many materials and techniques have been covered in brief with the time of the instructor considered foremost. This is just a starting place. I have given information on common problems and fixes but by no means have created an exhaustive compendium for solving all problems of 3D printing. I urge the reader to think in three areas: Is it hot? Does it move? Is the print adhering to the build plate? If the printer is not hot where it should be, break it down to connections, temperature and material. Is it plugged in? Is the temperature high or low enough? Are you using the right material for the project?

DOI: 10.4324/9781003376484-10

If it does not move, use the same ideas: Is it plugged in? Are they moving in the right direction? Is the tension of the belts correct?

If the print is not adhering to the plate: Is the nozzle close enough to the plate? Do you have the right "squish" factor? Are you using the right temperatures for the material used?

Considering safety, always read the SDS for the material you have and if you cannot find an SDS for the filament question whether you should let students use it. Always vent when you can and do not rely on your "judgment" to determine if fumes are safe or not. Move it up the chain to administration or safety. Always wear the personal protective equipment the manufacturer recommends.

The few projects laid out in this book should be considered as starting points. You, the reader, already have creativity and drive, confirmed because you are diving into the world of 3D printing, so trust your instincts. Have the student build interesting and challenging things. The challenge should not be making the 3D printer work, but in the execution of their designs. Start small and expand your students' knowledge as well as your own. If you only do the same projects repeatedly, you will not feed your creativity. This is an important resiliency thought to keep in mind. Teach, challenge and expand. I was once told that you should never peak but continually grow.

Grow yourself first. Your students will benefit from and learn how to grow themselves if you embody this idea in your teaching. Consider different angles, embrace failure and what you learn from it, change the rules and explore the world through experimentation and hope. Teach well.

11

GLOSSARY OF TERMS

ABS Acrylonitrile butadiene styrene.

Arduino A brand of programmable controller that has multiple inputs and outputs and is the base of many 3D printers.

Borosilicate glass A heat resistive glass that is used as a surface for many heat beds.

Bowden or pusher drive Extruder drive motor is located away from the hot end and uses a tube to guide the filament to the hot end.

Bowden tube A heat resistant plastic tube that guides filament.

Carbon fiber Usually PLA filament with carbon fiber mixed in. Will need a stainless-steel nozzle.

Cartesian x and y moving in one plane and z moving up and down from the print bed holding the hot end.

Core xy Build plate moves up and down instead of the extruder.

Cura Slicing program that has many features to help in 3D printing.

DOI: 10.4324/9781003376484-11

Direct drive An extruder that uses a non-geared drive to push filament.

FFF Fused filament fabrication.

Firmware Code that controls the controller and motors.

Fusion 360 Autodesk computer-aided design software specifically for 3D applications.

Heat bed Build plate that has temperature control to help in adhesion of the first layer.

Hot end A heat element for melting filament and a heat break to dissipate heat away from the heat block.

Kossel see Rostock.

Meshmixer 3D-sculpting software.

Nozzle Metal end of the hot end that will determine the diameter of the extruded filament.

Nylon Nylon-based filament.

PLA Polylactic acid.

Print head Heat block and nozzle component.

Prusa Josef Prusa refined a RepRap printer that now sports his name.

Repetier Open source 3D printer control software.

Rostock A three-tower z-axis controlled printer that uses placement of the extruder to control x and y.

Slic3r Slicing program that is easy to use and is included with many open source 3D printer control software.

Squish Compression of the first layer of the print to aid in adhesion.

Step stick Stepper motor control chip.

Stepper driver A controller that pulses the stepper motor to perform accurate motion.

Stepper motor A DC motor that has degree stops along its rotation.

Surface Outer shell of an item.

Tinkercad Free Autodesk computer-aided design software.

TPU Flexible filament.

Index

Note: Page numbers in *italic* refer to Figures.

3D camera 60
3D scanner 61, 71
3D scanning project 105

ABS (Acrylonitrile butadiene styrene) 15, 107; properties *13*
academic process, three stages of 27
adhesion 19–20, *79*, 90–93, 99, 118
agency 38
angles of detail 61–62
Anycubic 17
Arduino 2, 88, 120
Arduino Nano controller *88*
assembly 51, 52
assessment 24, 28; button project 63–64; candlestick phone project 54–55; honest 31; jewelry project 109; large print group project 96–97; lettered card project 34; makeup prosthesis project 77–78; science fiction communicator 103–104; self-assessment 31; student ownership of creation and 31

attempt, rigor based on 26–27
attributions license 49
Autodesk 35, 71, 121, 122
automotive double stick tape 92, *92*

baby wipes 73–74
batteries 86–87
bed, cleaning 19–20
bed y-axis 17
belts 3, 21
Blackburn, Barbara 27; *Rigor in Not a Four-Letter Word* 23
blame 32
Blender 71
Borosilicate glass 120
Bowden or pusher drive 120
Bowden tube 120
box tool 38, *39*, 99
brims 6
budget safety time (BST) triangle 16, *16*
build 4–6
build plate 53, 121
button project: additional rigor 65–68; assessment 63–64; "baking" the button 68; challenge areas in brief 68–69; guidelines and

rubric 63–64; model *66*;
painting or decorating 68;
problems and corrections
64–65; scanning 60–62,
64–65; selecting item 62;
shank design 66, *67*, 68;
testing 66–68

"CA" glue 92
cad program 60
candlestick phone project
48–59; additional rigor
56–57, *56*, *57*; assessment
54–55; challenge areas in
brief 59; debrief 55–56;
embody character trait 57;
guidelines and rubric
53–55; importing clean
version *52*; mirror tool *50*
carbon fiber 120
carbon filters 14
cartesian 120
centering jigs 114, *114*
chemical finishing 15
clamps/clamping 73, 94,
110–111, *111*
cleaning 19–20, *22*
coatings 15
Commedia masks 75, *76*
competition 27–28
compression engine 111
computer numerical control
(CNC) machine 60
constrained prompt 37, *37*
control card 2
core xy printer 17, 120
corncob 15
CPSG 3D scanner 64–65
Creality 17
Creation, student ownership
of assessment 31

creative common license
48–49
cube 1–2, *2*
Cura 4, 11, 120
cutting 93, *95*, 99, *99*, *100*, 115

DC motors 3, 81, 87, 122
debrief 38, 55–56
detail 61–62
direct drive 121
dowels 93; alignment *91*
drive train box *83*

electrical connections,
problem solving with
86–87
Etec 109
exporting 43–47, *44*
extruder 121
extruder drive motor 120
extruder feed 72–73, 74
extrusion recommendation
20–21

face of item *94*
failure: allowing for 31; and
learning 23–24
fasteners 21
feed cleaning screwdriver *22*
feedback loop 31
filament 12–13, 57,
117, 118; extrusion
recommendations 20–21;
flexible 71; temperature 4,
20–21, 71, 72–73, 74
finishing 15–16, 93
firmware 109, 121
fitment 107
flexible filament 71
foam clay 102
Formlabs 109

functional prints 9
fused filament fabrication 8, 9, 13, 17
Fusion 360 51, 71, 121

games 28
gantry x-axis 17
gears 81, *82*, 83, 85, 87–88
glue remover 73–74
grid infill 9–10, *10*
group vs individual projects 28–30
growth areas 36

hair ties 111
handles and hardware for antique furniture 106, *106*
heat: and extrusion 21; finishing 16; *see also* temperature
heat bed 20, 121
hem line measuring stand 112–113, *113*
HEPA filtration 14
hexagon infill 11, *11*
Hoover and Strong 108
hot end 121
hot forced air 93
how 25–26

individual vs group 28–30
infill 8–12
ink well and stand *105*

James, Duncan, *Antique Jewelry: Its Manufacture, Materials and Design* 108
jewelry project 107–109, *108*; assessment 109; pedagogical focus 108–109
journal 101

Kinect camera for Xbox 60
Kossel *see* Rostock

landing 53
large print group project 96–97
lazy Susan (rotating tray) 62
lettered card project 33–34, *39*; assessment 34; assets and objectives 35; debrief 38; exporting 43–47, *44*; faulty spacing *45*; overlapped spacing *46*; prompt 36–37, *36*, *37*; rubric 35–36; shape library tool *40*; sizing the box *41*; sizing the card to shapes *42*
LIDAR 61
line infill 11–12, *11*
linear actuator project 81–89; additional rigor 87–88; assembly *84*; assessment 85–86; challenge areas in brief 89; electrical connections 86–87; guidelines and rubric 85–86; Tinkercad gear creator *82*

magnet beds 20
Makerbot 17, 48
makeup prosthesis project 70–80; additional rigor 78; assessment 77–78; challenge areas in brief 80; filled nose 75; guidelines and rubric 76–78; nose, layers to increase adhesion *79*; painted nose *74*; safety 71, 74
masks 75–76

measuring aids 111–113, *113*
mechanical finishing 15
Mendel 17
Meshmixer 71, *72*, *72*, 78, 121
metal printing 109
motor driver 21
movement, printer 118
moving parts project *see* linear actuator project
multimeter 87
museum replica 104–107, *105*, *106*

Nema 14 and 11 3
Nema 23 and 34 3
Newman, Harold, *An Illustrated Dictionary of Jewelry* 108
non-commercial license 49
nose, scanning 71–72
nozzles 109, 118, 121
Nylon (polyamide) 121; properties *14*

open prompt 36–37, *36*
overkill factor 57

paint 15
painters' tape 6
painting or decorating *94*, 101, 105, 106–107
pendants 107–109
PETG (Glycolised polyester) 107; properties *14*
photo curing liquid 18
photogrammetry 60
PLA (polylactic acid) filament 8, 13–14, 74, 75, 90, 107, 120, 121; properties *13*
plastic 8–12

platform legs 115
polarity, reversing 87
polyurethane 90
prebuild description or drawing 31, 32
priming 93, *94*, 101
print head 121
printer: movement 118; selection 16–19
problems and corrections 19–21, *22*, 47, 117; button project 64–65; electrical connections 86–87; faulty spacing *45*; filament and temperature 72–73; landing 53; overlapped spacing *46*; scanning 64–65
process: and assessment 24, 28; caring about 23–24; rigor based on 26–27; supporting structures of 31; *see also* how
process lab, process of 30–32
prompts 35–37, *36*, *37*; reflective 101
Prusa 17, 121
Prusa clone printer 5

Qlone 61
questions to consider 115

raft 6, *6*
Raspberry Pi 2
razor scraper 19–20
ReconstructMe 61
Redwood, Ben, *The 3D Printing Handbook: Technologies, Design and Applications* 109
reference letters 93–94

reflective prompt 101
Repetier 121
RepRap 109, 121
resin cure printers 18
resolution 3–4, 65
reversing polarity 87
rigor: attempt and 26–27;
    button project 65–68;
    candlestick phone project
    55–56; linear actuator
    project 87–88; makeup
    prosthesis project 78
rings 107–109, *108*
Rio Grande 108
Rostock 18, 121
rubber bands 111

safe failing 24
safety data sheet (SDS) 14,
    71, 118
safety protocols 15, 18, 118;
    makeup prosthesis project
    71, 74
sanding 15
scanning: button project
    60–62, 64–65; nose 71–72;
    set up *62*
Scene Capture 61
science fiction communicator
    project 102–104; 3D
    printed *104*; assessment
    103–104; parts *103*
self-assessment 31
shape library tool *40*
shrink 51
Skanect 61
slic3r 4, 122
slicing 4–6
soldering 81, 87, 93
Soldiscape 109
spacing, faulty *45*

spirit gum 73
spool break 73, *73*
spring clamp/clip 73,
    110–111, *111*
squish 4, 5, 6, 19, *46*, 51, 118,
    122
stair tread measuring device
    111–112, *112*
steel wire rope cable strand *22*
step stick 122
stepper driver 122
stepper motors 2, *3*, 17, 18,
    88, 122
stereotyping 70
.stl file 43, 49, *72*
super glue 92
supports 6, *8*
surface 122
surgical tubing 112

taping 94
temperature 117; extrusion
    20; filament 4, 20–21, 71,
    72–73, 74; glue 92
test lamp function 87
text tool 38, *39*
thermoplastic polyurethane
    (TPU) 122
Thingiverse 48, 49, 111
threaded rods 3
Tinkercad 35, 48, 49, 51, 97,
    105, 122; gear creator *82*
tools to print 110–115
touch probe 60
TPU (thermoplastic
    polyurethane) 71
Triceratops skull 97–101, *98*,
    99, *100*
Trnio 61

Ultimaker 17

vaporising acetone 15
vibration, of working area
  94–96
volatile organic compounds
  (VOCs) 14

walking away 31
walnut shell 15
Watkins, Casey 76

waypoints 24
why of how 25–26

x-axis 2, 17

y-axis 2, 17
Yeggi 111

z-axis 2, 5, 5, 17, 18, 19, 51

Printed in the United States
by Baker & Taylor Publisher Services